Student Lab for Argument-Driven Inquiry in CHEMISTRY

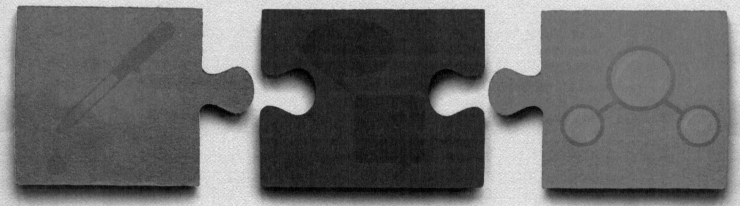

LAB INVESTIGATIONS for GRADES 9–12

Student Lab Manual for Argument-Driven Inquiry in CHEMISTRY

LAB INVESTIGATIONS for GRADES 9–12

Victor Sampson, Peter Carafano, Patrick Enderle, Steve Fannin, Jonathon Grooms, Sherry A. Southerland, Carol Stallworth, and Kiesha Williams

National Science Teachers Association
Arlington, Virginia

Claire Reinburg, Director
Wendy Rubin, Managing Editor
Rachel Ledbetter, Associate Editor
Amanda O'Brien, Associate Editor
Donna Yudkin, Book Acquisitions Coordinator

ART AND DESIGN
Will Thomas Jr., Director

PRINTING AND PRODUCTION
Catherine Lorrain, Director

NATIONAL SCIENCE TEACHERS ASSOCIATION
David L. Evans, Executive Director
David Beacom, Publisher

1840 Wilson Blvd., Arlington, VA 22201
www.nsta.org/store
For customer service inquiries, please call 800-277-5300.

Copyright © 2016 by the National Science Teachers Association.
All rights reserved. Printed in the United States of America.
19 18 17 16   5 4 3 2

NSTA is committed to publishing material that promotes the best in inquiry-based science education. However, conditions of actual use may vary, and the safety procedures and practices described in this book are intended to serve only as a guide. Additional precautionary measures may be required. NSTA and the authors do not warrant or represent that the procedures and practices in this book meet any safety code or standard of federal, state, or local regulations. NSTA and the authors disclaim any liability for personal injury or damage to property arising out of or relating to the use of this book, including any of the recommendations, instructions, or materials contained therein.

PERMISSIONS
Book purchasers may photocopy, print, or e-mail up to five copies of an NSTA book chapter for personal use only; this does not include display or promotional use. Elementary, middle, and high school teachers may reproduce forms, sample documents, and single NSTA book chapters needed for classroom or noncommercial, professional-development use only. E-book buyers may download files to multiple personal devices but are prohibited from posting the files to third-party servers or websites, or from passing files to non-buyers. For additional permission to photocopy or use material electronically from this NSTA Press book, please contact the Copyright Clearance Center (CCC) (*www.copyright.com*; 978-750-8400). Please access *www.nsta.org/permissions* for further information about NSTA's rights and permissions policies.

Cataloging-in-Publication Data are available from the Library of Congress.
  LCCN: 2014029558

CONTENTS

Acknowledgments ix
About the Authors xi

SECTION 1
Introduction and Lab Safety

Introduction by Victor Sampson 3

Safety in the Science Classroom, Laboratory, or Field Sites 5

SECTION 2—Physical Sciences Core Idea 1.A
Structure and Properties of Matter

INTRODUCTION LABS

Lab 1. Bond Character and Molecular Polarity: How Does Atom Electronegativity Affect Bond Character and Molecular Polarity?
- Lab Handout 14
- Checkout Questions 19

Lab 2. Molecular Shapes: How Does the Number of Substituents Around a Central Atom Affect the Shape of a Molecule?
- Lab Handout 21
- Checkout Questions 25

Lab 3. Rate of Dissolution: Why Do the Surface Area of the Solute, the Temperature of the Solvent, and the Amount of Agitation That Occurs When the Solute and the Solvent Are Mixed Affect the Rate of Dissolution?
- Lab Handout 29
- Checkout Questions 34

Lab 4. Molarity: What Is the Mathematical Relationship Between the Moles of a Solute, the Volume of the Solvent, and the Molarity of an Aqueous Solution?
- Lab Handout 36
- Checkout Questions 41

Lab 5. Temperature Changes Due to Evaporation: Which of the Available Substances Has the Strongest Intermolecular Forces?
- Lab Handout 43
- Checkout Questions 49

Lab 6. Pressure, Temperature, and Volume of Gases: How Does Changing the Volume or Temperature of a Gas Affect the Pressure of That Gas?
- Lab Handout 51
- Checkout Questions 57

Lab 7. Periodic Trends: Which Properties of the Elements Follow a Periodic Trend?
    Lab Handout 59
    Checkout Questions 63

Lab 8. Solutes and the Freezing Point of Water: How Does the Addition of Different Types of Solutes Affect the Freezing Point of Water?
    Lab Handout 65
    Checkout Questions 70

APPLICATION LABS

Lab 9. Melting and Freezing Points: Why Do Substances Have Specific Melting and Freezing Points?
    Lab Handout 74
    Checkout Questions 80

Lab 10. Identification of an Unknown Based on Physical Properties: What Type of Solution Is the Unknown Liquid?
    Lab Handout 83
    Checkout Questions 88

Lab 11. Atomic Structure and Electromagnetic Radiation: What Are the Identities of the Unknown Powders?
    Lab Handout 91
    Checkout Questions 96

Lab 12. Magnetism and Atomic Structure: What Relationships Exist Between the Electrons in a Substance and the Strength of Magnetic Attraction?
    Lab Handout 98
    Checkout Questions 104

Lab 13. Density and the Periodic Table: What Are the Densities of Germanium and Flerovium?
    Lab Handout 106
    Checkout Questions 111

Lab 14. Molar Relationships: What Are the Identities of the Unknown Compounds?
    Lab Handout 114
    Checkout Questions 118

Lab 15. The Ideal Gas Law: How Can a Value of R for the Ideal Gas Law Be Accurately Determined Inside the Laboratory?
    Lab Handout 120
    Checkout Questions 126

SECTION 3— Physical Sciences Core Idea 1.B
Chemical Reactions

INTRODUCTION LABS

Lab 16. Development of a Reaction Matrix: What Are the Identities of the Unknown Chemicals?
- Lab Handout 134
- Checkout Questions 139

Lab 17. Limiting Reactants: Why Does Mixing Reactants in Different Mole Ratios Affect the Amount of the Product and the Amount of Each Reactant That Is Left Over?
- Lab Handout 142
- Checkout Questions 149

Lab 18. Characteristics of Acids and Bases: How Can the Chemical Properties of an Aqueous Solution Be Used to Identify It as an Acid or a Base?
- Lab Handout 152
- Checkout Questions 157

Lab 19. Strong and Weak Acids: Why Do Strong and Weak Acids Behave in Different Manners Even Though They Have the Same Chemical Properties?
- Lab Handout 159
- Checkout Questions 165

Lab 20. Enthalpy Change of Solution: How Can Chemists Use the Properties of a Solute to Predict If an Enthalpy Change of Solution Will Be Exothermic or Endothermic?
- Lab Handout 167
- Checkout Questions 174

Lab 21. Reaction Rates: Why Do Changes in Temperature and Reactant Concentration Affect the Rate of a Reaction?
- Lab Handout 177
- Checkout Questions 183

Lab 22. Chemical Equilibrium: Why Do Changes in Temperature, Reactant Concentration, and Product Concentration Affect the Equilibrium Point of a Reaction?
- Lab Handout 186
- Checkout Questions 193

APPLICATION LABS

Lab 23. Classification of Changes in Matter: Which Changes Are Examples of a Chemical Change, and Which Are Examples of a Physical Change?
Lab Handout 198
Checkout Questions 202

Lab 24. Identification of Reaction Products: What Are the Products of the Chemical Reactions?
Lab Handout 205
Checkout Questions 210

Lab 25. Acid-Base Titration and Neutralization Reactions: What Is the Concentration of Acetic Acid in Each Sample of Vinegar?
Lab Handout 212
Reference Sheet 219
Checkout Questions 220

Lab 26. Composition of Chemical Compounds: What Is the Empirical Formula of Magnesium Oxide?
Lab Handout 222
Checkout Questions 227

Lab 27. Stoichiometry and Chemical Reactions: Which Balanced Chemical Equation Best Represents the Thermal Decomposition of Sodium Bicarbonate?
Lab Handout 229
Checkout Questions 235

Lab 28. Designing a Cold Pack: Which Salt Should Be Used to Make an Effective but Economical Cold Pack?
Lab Handout 237
Checkout Questions 242

Lab 29. Rate Laws: What Is the Rate Law for the Reaction Between Hydrochloric Acid and Sodium Thiosulfate?
Lab Handout 245
Reference Sheet 251
Checkout Questions 252

Lab 30. Equilibrium Constant and Temperature: How Does a Change in Temperature Affect the Value of the Equilibrium Constant for an Exothermic Reaction?
Lab Handout 255
Checkout Questions 262

Image Credits 265

ACKNOWLEDGMENTS

The development of this book was supported by the Institute of Education Sciences, U.S. Department of Education, through grant R305A100909 to Florida State University. The opinions expressed are those of the authors and do not represent the views of the institute or the U.S. Department of Education.

ABOUT THE AUTHORS

Victor Sampson is an associate professor of STEM education and the director of the Center for STEM Education at The University of Texas at Austin (UT-Austin). He received a BA in zoology from the University of Washington, an MIT from Seattle University, and a PhD in curriculum and instruction with a specialization in science education from Arizona State University. Victor also taught high school biology and chemistry for nine years. He specializes in argumentation in science education, teacher learning, and assessment. To learn more about his work in science education, go to *www.vicsampson.com*.

Peter Carafano is an associate professor in the Science Department at Florida State University Schools, the K–12 research lab school for Florida State University (FSU), where he teaches high school and AP chemistry as well as health science. He began his career in education while working as a paramedic for the Delray Beach (FL) Fire Department. As a specialist in hazardous materials management and pre-hospital medical toxicology, Peter became a visiting instructor, lecturer, and author on hazardous materials and fire science chemistry. After retiring from the fire service in 2000, he obtained an MEd at FSU and began a new career as a teacher. Peter is also the director of the State of Florida Student Astronaut Challenge competition held at the Kennedy Space Center each year and the managing director of *The Journal of Emergency Medical Service Responders*, a student-driven and professionally peer-reviewed magazine.

Patrick Enderle is an assistant professor of science education in the Department of Middle and Secondary Education at Georgia State University. He received his BS and MS in molecular biology from East Carolina University. Patrick spent some time as a high school biology teacher and several years as a visiting professor in the Department of Biology at East Carolina University. He then attended FSU, where he graduated with a PhD in science education. His research interests include argumentation in the science classroom, science teacher professional development, and enhancing undergraduate science education.

Steve Fannin is an assistant in research at the Florida Center for Research in Science, Technology, Engineering, and Mathematics (FCR-STEM) at FSU. Before taking his current position, he was a high school science teacher for 31 years. Steve received a BS in biological science from FSU and is a National Board Certified Teacher. He was honored with the Presidential Award for Excellence in Mathematics and Science Teaching in 2011.

Jonathon Grooms is an assistant professor of curriculum and pedagogy in the Graduate School of Education and Human Development at The George Washington University. He received a BS in secondary science and mathematics teaching with a focus in chemistry and physics from FSU. Upon graduation, Jonathon joined FSU's

ABOUT THE AUTHORS

Office of Science Teaching, where he directed the physical science outreach program, Science on the Move. He also earned a PhD in science education from FSU.

Sherry A. Southerland is a professor at FSU and the co-director of FSU-Teach. FSU-Teach is a collaborative math and science teacher preparation program between the College of Arts and Sciences and the College of Education. She received a BS and an MS in biology from Auburn University and a PhD in curriculum and instruction from Louisiana State University, with a specialization in science education and evolutionary biology. Sherry has worked as a teacher educator, biology instructor, high school science teacher, field biologist, and forensic chemist. Her research interests include understanding the influence of culture and emotions on learning—specifically evolution education and teacher education—and understanding how to better support teachers in shaping the way they approach science teaching and learning.

Carol Stallworth is an honors and AP chemistry teacher at Lincoln High School in Tallahassee, Florida, and an adjunct chemistry professor at Tallahassee Community College. She received a BS and an MS in biochemistry from FSU. Carol has conducted biochemical research and wrote the article, "Cooperativity in Monomeric Enzymes With Single-Ligand Binding Sites," which appeared in the journal *Bioorganic Chemistry* in 2011.

Kiesha Williams is a chemistry teacher at Cypress Woods High School in Cypress, Texas. Before taking her current position, she taught chemistry at Florida State University Schools in Tallahassee, FL. Kiesha received a BS in chemistry and Spanish from the University of South Dakota and an MEd in science education from FSU. She has published two articles in *The Science Teacher*.

SECTION 1
Introduction and Lab Safety

INTRODUCTION

by Victor Sampson

Science is much more than a body of knowledge or a set of core ideas that reflect our current understanding of how the world works and why it works that way. Science is also a set of crosscutting concepts and practices that people can use to develop and refine new explanations for, or descriptions of, the natural world. These core ideas, crosscutting concepts, and practices of science are important for you to learn. When you understand these, it is easier to appreciate the beauty and wonder of science, to engage in public discussions about science, and to critique the merits of scientific findings that are presented through the popular media. You will also have the knowledge and skills needed to continue learning about science outside school or to enter a career in science, engineering, or technology.

The core ideas of science include the theories, laws, and models that scientists use to explain natural phenomena and bodies of data and to predict the outcomes of new investigations. The crosscutting concepts are themes that have value in every discipline of science and are used to help us understand a natural phenomenon. They can be used as organizational frameworks for connecting knowledge from the various fields of science into a coherent and scientifically based view of the world. Finally, the practices of science are used to develop and refine new ideas about the world. Although some practices differ from one field of science to another, all fields share a set of common practices. The practices include such things as asking and answering questions; planning and carrying out investigations; analyzing and interpreting data; and obtaining, evaluating, and communicating information. One of the most important practices of science is arguing from evidence. Arguing from evidence, or the practice of proposing, supporting, challenging, and refining claims based on evidence, is important because scientists need to be able to examine, review, and evaluate their own ideas and to critique those of others. Scientists also argue from evidence when they need to appraise the quality of data, produce and improve models, develop new testable questions from those models, and suggest ways to refine or modify existing theories, laws, and models.

It is important to always remember that science is a social activity, not an individual one. Science is social because many different scientists contribute to the development of new scientific knowledge. As scientists carry out their research, they frequently talk with their colleagues, both formally and informally. They exchange emails, engage in discussions at conferences, share research techniques and analytical procedures, and present new ideas by writing articles in journals or chapters in books. They also critique the ideas and methods used by other scientists through a formal peer review process before they can be published in journals or books. In short, scientists are members of a community, the members of which work together to build, develop, test, critique, and refine ideas. The ways scientists talk, write, think, and interact with each other reflect common ideas about what

INTRODUCTION

counts as quality and shared standards for how new ideas should be developed, shared, evaluated, and refined. These ways of interacting make science different from other ways of knowing. The core ideas, crosscutting concepts, and practices of science are important within the scientific community because most, if not all, members of that community find them to be a useful way to develop and refine new explanations for, or descriptions of, the natural world.

The laboratory investigations included in this book are designed to help you learn the core ideas, crosscutting concepts, and practices of science. During each investigation, you will have an opportunity to use a core idea, several crosscutting concepts, and the practices of science to understand a natural phenomenon or solve a problem. Your teacher will introduce each investigation by giving you a task to accomplish and a guiding question to answer. You will then work as part of a team to plan and carry out an investigation to collect the data you need to answer that question. From there, your team will develop an initial argument that includes a claim, evidence in support of your claim, and a justification of your evidence. The claim will be your answer to the guiding question, the evidence will include your analysis of the data you collected and an interpretation of that analysis, and the justification will explain why your evidence is important. Next, you will have an opportunity to share your argument with your classmates and to critique their arguments, much like professional scientists do. You will then revise your initial argument based on your colleagues' feedback. Finally, you will be asked to write an investigation report on your own to share what you learned. The report will go through double-blind peer review so you can improve it before you submit it to you teacher for a grade. As you complete more and more investigations in this lab manual, you will not only learn the core ideas associated with each investigation but also get better at using the crosscutting concepts and practices of science to understand the natural world.

SAFETY IN THE SCIENCE CLASSROOM, LABORATORY, OR FIELD SITES

***Note to science teachers and supervisors/administrators:** The following safety acknowledgment form is for your use in the classroom and should be given to students at the beginning of the school year to help them understand their role in ensuring a safer and productive science experience.*

Science is a process of discovering and exploring the natural world. Exploration occurs in the classroom/laboratory or in the field. As part of your science class, you will be doing many activities and investigations that will involve the use of various materials, equipment, and chemicals. Safety in the science classroom, laboratory, or field sites is the FIRST PRIORITY for students, instructors, and parents. To ensure safer classroom/laboratory/field experiences, the following **Science Rules and Regulations** have been developed for the protection and safety of all. Your instructor will provide additional rules for specific situations or settings. The rules and regulations must be followed at all times. After you have reviewed them with your instructor, read and review the rules and regulations with your parent/guardian. Their signature and your signature on the safety acknowledgment form are required before you will be permitted to participate in any activities or investigations. Your signature indicates that you have read these rules and regulations, understand them, and agree to follow them at all times while working in the classroom/laboratory or in the field.

Source: National Science Teachers Association (NSTA). Safety in the Science Classroom. *www.nsta.org/pdfs/SafetyInTheScienceClassroom.pdf.*

Safety in the Science Classroom, Laboratory, or Field Sites

Safety Standards of Student Conduct in the Classroom, Laboratory, and in the Field

1. Conduct yourself in a responsible manner at all times. Frivolous activities, mischievous behavior, throwing items, and conducting pranks are prohibited.

2. Lab and safety information and procedures must be read ahead of time. All verbal and written instructions shall be followed in carrying out the activity or investigation.

3. Eating, drinking, gum chewing, applying cosmetics, manipulating contact lenses, and other unsafe activities are not permitted in the laboratory.

4. Working in the laboratory without the instructor present is prohibited.

5. Unauthorized activities or investigations are prohibited. Unsupervised work is not permitted.

6. Entering preparation or chemical storage areas is prohibited at all times.

7. Removing chemicals or equipment from the classroom or laboratory is prohibited unless authorized by the instructor.

Personal Safety

8. Sanitized indirectly vented chemical splash goggles or safety glasses as appropriate (meeting the ANSI Z87.1 standard) shall be worn during activities or demonstrations in the classroom, laboratory, or field, including pre-laboratory work and clean-up, unless the instructor specifically states that the activity or demonstration does not require the use of eye protection.

9. When an activity requires the use of laboratory aprons, the apron shall be appropriate to the size of the student and the hazard associated with the activity or investigation. The apron shall remain tied throughout the activity or investigation.

10. All accidents, chemical spills, and injuries must be reported immediately to the instructor, no matter how trivial they may seem at the time. Follow your instructor's directions for immediate treatment.

11. Dress appropriately for laboratory work by protecting your body with clothing and shoes. This means that you should use hair ties to tie back long hair and tuck into the collar. Do not wear loose or baggy clothing or dangling jewelry on laboratory days. Acrylic nails are also a safety hazard near heat sources and should not be used. Sandals or open-toed shoes are not to be worn during any lab activities. Refer to pre-lab instructions. If in doubt, ask!

Safety in the Science Classrom, Laboratory, or Field Sites

12. Know the location of all safety equipment in the room. This includes eye wash stations, the deluge shower, fire extinguishers, the fume hood, and the safety blanket. Know the location of emergency master electric and gas shut offs and exits.

13. Certain classrooms may have living organisms including plants in aquaria or other containers. Students must not handle organisms without specific instructor authorization. Wash your hands with soap and water after handling organisms and plants.

14. When an activity or investigation requires the use of laboratory gloves for hand protection, the gloves shall be appropriate for the hazard and worn throughout the activity.

Specific Safety Precautions Involving Chemicals and Lab Equipment

15. Avoid inhaling in fumes that may be generated during an activity or investigation.

16. Never fill pipettes by mouth suction. Always use the suction bulbs or pumps.

17. Do not force glass tubing into rubber stoppers. Use glycerin as a lubricant and hold the tubing with a towel as you ease the glass into the stopper.

18. Proper procedures shall be followed when using any heating or flame producing device especially gas burners. Never leave a flame unattended.

19. Remember that hot glass looks the same as cold glass. After heating, glass remains hot for a very long time. Determine if an object is hot by placing your hand close to the object but do not touch it.

20. Should a fire drill, lockdown, or other emergency occur during an investigation or activity, make sure you turn off all gas burners and electrical equipment. During an evacuation emergency, exit the room as directed. During a lockdown, move out of the line of sight from doors and windows if possible or as directed.

21. Always read the reagent bottle labels twice before you use the reagent. Be certain the chemical you use is the correct one.

22. Replace the top on any reagent bottle as soon as you have finished using it and return the reagent to the designated location.

23. Do not return unused chemicals to the reagent container. Follow the instructor's directions for the storage or disposal of these materials.

Safety in the Science Classrom, Laboratory, or Field Sites

Standards For Maintaining a Safer Laboratory Environment

24. Backpacks and books are to remain in an area designated by the instructor and shall not be brought into the laboratory area.

25. Never sit on laboratory tables.

26. Work areas should be kept clean and neat at all times. Work surfaces are to be cleaned at the end of each laboratory or activity.

27. Solid chemicals, metals, matches, filter papers, broken glass, and other materials designated by the instructor are to be deposited in the proper waste containers, not in the sink. Follow your instructor's directions for disposal of waste.

28. Sinks are to be used for the disposal of water and those solutions designated by the instructor. Other solutions must be placed in the designated waste disposal containers.

29. Glassware is to be washed with hot, soapy water and scrubbed with the appropriate type and sized brush, rinsed, dried, and returned to its original location.

30. Goggles are to be worn during the activity or investigation, clean up, and through hand washing.

31. Safety Data Sheets (SDSs) contain critical information about hazardous chemicals of which students need to be aware. Your instructor will review the salient points on the SDSs for the hazardous chemicals students will be working with and also post the SDSs in the lab for future reference.

Safety in the Science Classroom, Laboratory, or Field Sites

Safety Acknowledgment Form: Science Rules and Regulations

I have read the science rules and regulations in the *Student Lab Manual for Argument-Driven Inquiry in Chemistry*, and I agree to follow them during any science course, investigation, or activity. By signing this form, I acknowledge that the science classroom, laboratory, or field sites can be an unsafe place to work and learn. The safety rules and regulations are developed to help prevent accidents and to ensure my own safety and the safety of my fellow students. I will follow any additional instructions given by my instructor. I understand that I may ask my instructor at any time about the rules and regulations if they are not clear to me. My failure to follow these science laboratory rules and regulations may result in disciplinary action.

_____   _____

Student Signature   Date

_____   _____

Parent/Guardian Signature   Date

SECTION 2
Physical Sciences Core Idea 1.A

Structure and Properties of Matter

Introduction Labs

LAB 1

Lab Handout

Lab 1. Bond Character and Molecular Polarity: How Does Atom Electronegativity Affect Bond Character and Molecular Polarity?

Introduction

Chemists often classify chemical compounds into one of two broad categories. The first category is molecular compounds, and the second category is ionic compounds. Molecular compounds consist of atoms that are held together by covalent bonds. Ionic compounds, in contrast, are composed of positive and negative ions that are joined by ionic bonds. Covalent bonds are formed when atoms share one or more pairs of electrons. An ionic bond is formed when one or more electrons from one atom are transferred to another atom. The transfer of one or more electrons from one atom to another results in the formation of a positive ion and a negative ion. The ions then attract each other because they have opposite electrical charges.

The term *electronegativity* refers to a measure of an atom's tendency to attract electrons from other atoms. Atom electronegativity affects the nature or the character of the bond that will form between two atoms. The electronegativity of atoms also affects the electrical charge of a molecular compound. In some molecules, the electronegativity of the atoms that make up the molecule results in one side of the molecule having a partial negative electrical charge and the other side having a partial positive charge. When this happens, the molecule is described as being polar. Water is an example of a polar molecule because the oxygen side of the molecule has a partial negative charge and the hydrogen side of the molecule has a partial positive charge. Nonpolar molecules, in contrast, do not have electrical poles. Carbon dioxide is an example of a nonpolar molecule because both sides of the molecule have the same charge.

In this investigation, you will explore the relationship between the electronegativity of the atoms found within a chemical compound and the character of the bond that holds that compound together. You will also explore how atom electronegativity and molecular polarity are related.

Your Task

Use a computer simulation to explore the effect of atom electronegativity on bond character and molecular polarity.

Bond Character and Molecular Polarity
How Does Atom Electronegativity Affect Bond Character and Molecular Polarity?

The guiding question of this investigation is, **How does atom electronegativity affect bond character and molecular polarity?**

Materials
You will use an online simulation called *Molecule Polarity* to conduct your investigation. You can access the simulation by going to the following website: *http://phet.colorado.edu/en/simulation/molecule-polarity*.

Safety Precautions
Follow all normal lab safety rules.

Investigation Proposal Required?   ☐ Yes   ☐ No

Getting Started
The *Molecule Polarity* simulation (see Figure L1.1, p. 16) enables you to create molecules with different numbers of atoms in them and to adjust the electronegativity of each atom in the molecule. You can also view the partial charge of each side of the molecule, the electrostatic potential across the molecule, and the bond character. This information will allow you to explore how atom electronegativity affects bond character and molecular polarity.

To configure the simulation for this investigation, click on "Bond Character" in the View box and on "Electrostatic Potential" in the Surface box. This will allow you to explore how changing the electronegativity of atoms affects the nature of the bond that forms between them. It will also allow you to examine how atom electronegativity affects the electrical charge of a chemical compound. The other options, such as "Bond Dipole" and "Partial Charges" in the View box and "None" and "Electron Density" in the Surface box, should not be checked. Once the simulation is ready to use, you must determine what type of data you will need to collect, how you will collect the data, and how you will analyze the data to answer the guiding question.

To determine *what type of data you need to collect*, think about the following questions:

- What type of observations will you need to record during your investigation?
- When will you need to make these observations?

To determine *how you will collect the data*, think about the following questions:

- What types of molecules will you need to include in the simulation (i.e., molecules made up of two atoms, molecules made up of three atoms, or both)?
- What range of electronegativity values will you need to investigate?
- What types of comparisons will you need to make?

LAB 1

FIGURE L1.1

A screenshot of the *Molecule Polarity* simulation

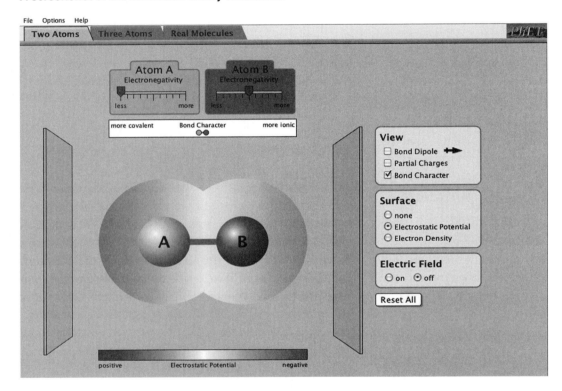

- How will you keep track of the data you collect and how will you organize it?

To determine *how you will analyze the data*, think about the following questions:

- What type of calculations will you need to make?
- What type of graph could you create to help make sense of your data?

Connections to Crosscutting Concepts, the Nature of Science, and the Nature of Scientific Inquiry

As you work through your investigation, be sure to think about

- the importance of looking for and identifying patterns,
- how models are used to study natural phenomena,
- how the structure of an object is related to its function,
- the difference between laws and theories in science, and
- the difference between data and evidence in science.

Bond Character and Molecular Polarity
How Does Atom Electronegativity Affect Bond Character and Molecular Polarity?

Initial Argument

Once your group has finished collecting and analyzing your data, you will need to develop an initial argument. Your argument must include a *claim*, which is your answer to the guiding question. Your argument must also include *evidence* in support of your claim. The evidence is your analysis of the data and your interpretation of what the analysis means. Finally, you must include a *justification* of the evidence in your argument. You will therefore need to use a scientific concept or principle to explain why the evidence that you decided to use is relevant and important. You will create your initial argument on a whiteboard. Your whiteboard must include all the information shown in Figure L1.2.

FIGURE L1.2
Argument presentation on a whiteboard

The Guiding Question:	
Our Claim:	
Our Evidence:	Our Justification of the Evidence:

Argumentation Session

The argumentation session allows all of the groups to share their arguments. One member of each group stays at the lab station to share that group's argument, while the other members of the group go to the other lab stations one at a time to listen to and critique the arguments developed by their classmates. The goal of the argumentation session is not to convince others that your argument is the best one; rather, the goal is to identify errors or instances of faulty reasoning in the initial arguments so these mistakes can be fixed. You will therefore need to evaluate the content of the claim, the quality of the evidence used to support the claim, and the strength of the justification of the evidence included in each argument that you see. To critique an argument, you might need more information than what is included on the whiteboard. You might therefore need to ask the presenter one or more follow-up questions, such as:

- What did your group do to analyze the data, and why did you decide to do it that way?
- Is that the only way to interpret the results of your group's analysis? How do you know that your interpretation of the analysis is appropriate?
- Why did your group decide to present your evidence in that manner?
- What other claims did your group discuss before deciding on that one? Why did you abandon those alternative ideas?
- How confident are you that your group's claim is valid? What could you do to increase your confidence?

Once the argumentation session is complete, you will have a chance to meet with your group and revise your original argument. Your group might need to gather more data or design a way to test one or more alternative claims as part of this process. Remember, your

LAB 1

goal at this stage of the investigation is to develop the most valid or acceptable answer to the research question!

Report

Once you have completed your research, you will need to prepare an *investigation report* that consists of three sections that provide answers to the following questions:

1. What question were you trying to answer and why?
2. What did you do during your investigation and why did you conduct your investigation in this way?
3. What is your argument?

Your report should answer these questions in two pages or less. The report must be typed and any diagrams, figures, or tables should be embedded into the document. Be sure to write in a persuasive style; you are trying to convince others that your claim is acceptable or valid!

Checkout Questions

Lab 1. Bond Character and Molecular Polarity: How Does Atom Electronegativity Affect Bond Character and Molecular Polarity?

1. How does the electronegativity of atoms affect the polarity of a molecule?

2. Petroleum products are used for a variety of applications, including fueling automobiles, cooking, and heating homes. Petroleum is composed of a variety of molecules containing different numbers of carbon and hydrogen atoms bonded together. On several occasions, large amounts of petroleum have spilled into the ocean from shipwrecks. When the petroleum spills into the ocean it floats because it is less dense than water, but it also does not mix with the water due to the nature of its chemical bonds.

   Use what you know about bond types and electronegativity to explain why the petroleum molecules do not mix with water molecules.

3. The terms *data* and *evidence* have the same meaning in science.

   a. I agree with this statement
   b. I disagree with this statement

   Explain your answer, using an example from your investigation about bond character and molecular polarity.

LAB 1

4. Theories can become laws over time.

    a. I agree with this statement
    b. I disagree with this statement

    Explain your answer, using an example from your investigation about bond character and molecular polarity.

5. Scientists often use models to help them understand natural phenomena. Explain what a model is and why models are important, using an example from your investigation about bond character and molecular polarity.

6. Scientists often look for and attempt to explain patterns in nature. Explain why patterns are important, using an example from your investigation about bond character and molecular polarity.

Lab Handout

Lab 2. Molecular Shapes: How Does the Number of Substituents Around a Central Atom Affect the Shape of a Molecule?

Introduction

Molecules are three-dimensional entities and therefore should be depicted in three dimensions. We can translate the two-dimensional electron dot structure representing a molecule into a more useful three-dimensional rendering by using a tool known as the valance shell electron pair repulsion (VSEPR) model. According to this model, any given pair of valance shell electrons strives to get as far away as possible from all other electron pairs in the shell. This includes both nonbonding pairs of electrons and any bonding pair not taking part in a double or triple bond. Pairs of electrons in a multiple bond stay together because of their mutual attractions for the same two nuclei. It is this striving for maximum separation distance between electron pairs that determines the geometry of any molecule.

The VSEPR model allows us to use the electron dot structures of atoms to predict the three-dimensional geometry of simple molecules. This geometry is determined by considering the number of *substituents* surrounding a central atom. A substituent is any bonded atom or nonbonding pair of electrons. For example, the carbon on the methane molecule (CH_4) shown in Figure L2.1 has four substituents (four hydrogen atoms). The oxygen atom of water also has four substituents (two hydrogen atoms and two nonbonding pairs of electrons).

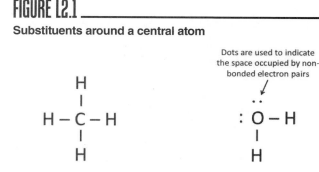

FIGURE L2.1
Substituents around a central atom

Your Task

Use a computer simulation to develop a rule that you can use to predict the shape of a molecule based on the number of atoms and lone electron pairs (i.e., substituents) around a central atom.

LAB 2

The guiding question of this investigation is, **How does the number of substituents around a central atom affect the shape of a molecule?**

Materials

You will use an online simulation called *Molecule Shapes* to conduct your investigation. You can access the simulation by going to the following website: *http://phet.colorado.edu/en/simulation/molecule-shapes*.

Safety Precautions

Follow all normal lab safety rules.

Investigation Proposal Required?   ☐ Yes   ☐ No

Getting Started

The *Molecule Shapes* simulation (see Figure L2.2) models how the number of atoms and lone electron pairs around a central atom affects the shape of a molecule. With this simulation, you can decide how many atoms are found in the molecule, the nature of the bonds found between the atoms (single, double, or triple) and the number of lone electron pairs that are found around the central atom. Your goal is to develop a rule that you can use to predict the shape of a molecule based on the number of atoms and lone electron pairs (i.e., substituents) found around the central atom. Before you start using the simulation, however, you must determine what type of data you will need to collect, how you will collect the data, and how you will analyze the data to answer the guiding question.

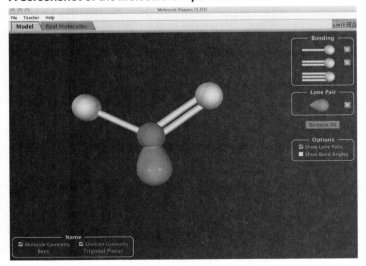

FIGURE L2.2
A screenshot of the *Molecule Shapes* simulation

To determine *what type of data you need to collect*, think about the following questions:

- What type of observations will you need to record during your investigation?
- When will you need to make these observations?

To determine *how you will collect the data*, think about the following questions:

Molecular Shapes

How Does the Number of Substituents Around a Central Atom Affect the Shape of a Molecule?

- What types of molecules will you need to examine using the simulation?
- What type of comparisons will you need to make?
- How will you keep track of the data you collect and how will you organize it?

To determine *how you will analyze the data*, think about the following questions:

- What type of calculations will you need to make?
- What type of graph could you create to help make sense of your data?

Connections to Crosscutting Concepts, the Nature of Science, and the Nature of Scientific Inquiry

As you work through your investigation, be sure to think about

- the importance of looking for and identifying patterns,
- how models are used to study natural phenomena,
- how the structure of an object is related to its function,
- the difference between laws and theories in science, and
- the wide range of methods that can be used by scientists during an investigation.

Initial Argument

Once your group has finished collecting and analyzing your data, you will need to develop an initial argument. Your argument must include a *claim*, which is your answer to the guiding question. Your argument must also include *evidence* in support of your claim. The evidence is your analysis of the data and your interpretation of what the analysis means. Finally, you must include a *justification* of the evidence in your argument. You will therefore need to use a scientific concept or principle to explain why the evidence that you decided to use is relevant and important. You will create your initial argument on a whiteboard. Your whiteboard must include all the information shown in Figure L2.3.

FIGURE L2.3
Argument presentation on a whiteboard

The Guiding Question:	
Our Claim:	
Our Evidence:	Our Justification of the Evidence:

Argumentation Session

The argumentation session allows all of the groups to share their arguments. One member of each group stays at the lab station to share that group's argument, while the other members of the group go to the other lab stations one at a time to listen to and critique the arguments developed by their classmates. The goal of the argumentation session is not to convince others that your argument is the best one; rather, the goal is to identify errors or instances of faulty reasoning in the initial arguments so these mistakes can be fixed. You will therefore need to evaluate the content of the claim, the quality of the evidence used to support the claim, and the strength of the justification of the evidence included in each argument that you see. To critique an argument, you might need more information than what is included on the whiteboard. You might therefore need to ask the presenter one or more follow-up questions, such as:

- What did your group do to analyze the data, and why did you decide to do it that way?
- Is that the only way to interpret the results of your group's analysis? How do you know that your interpretation of the analysis is appropriate?
- Why did your group decide to present your evidence in that manner?
- What other claims did your group discuss before deciding on that one? Why did you abandon those alternative ideas?
- How confident are you that your group's claim is valid? What could you do to increase your confidence?

Once the argumentation session is complete, you will have a chance to meet with your group and revise your original argument. Your group might need to gather more data or design a way to test one or more alternative claims as part of this process. Remember, your goal at this stage of the investigation is to develop the most valid or acceptable answer to the research question!

Report

Once you have completed your research, you will need to prepare an *investigation report* that consists of three sections that provide answers to the following questions:

1. What question were you trying to answer and why?
2. What did you do during your investigation and why did you conduct your investigation in this way?
3. What is your argument?

Your report should answer these questions in two pages or less. The report must be typed and any diagrams, figures, or tables should be embedded into the document. Be sure to write in a persuasive style; you are trying to convince others that your claim is acceptable or valid!

Checkout Questions

Lab 2. Molecular Shapes: How Does the Number of Substituents Around a Central Atom Affect the Shape of a Molecule?

1. Describe the basic principle of the valance shell electron pair repulsion (VSEPR) model.

   Use what you know about the VSEPR model to answer questions 2 and 3.

2. Draw the Lewis dot structure for CO_2 in the space below.

LAB 2

What shape is a molecule of CO_2?

    a. Linear

    b. Trigonal planar

    c. Bent

    d. Tetrahedral

Explain your answer.

3. Draw the Lewis dot structure for CCl_4 in the space below.

What shape is a molecule of CCl_4?

    a. Linear

    b. Trigonal planar

    c. Bent

    d. Tetrahedral

Explain your answer.

Molecular Shapes
How Does the Number of Substituents Around a Central Atom Affect the Shape of a Molecule?

4. All scientists follow the scientific method during an investigation.

    a. I agree with this statement.
    b. I disagree with this statement.

    Explain your answer, using an example from your investigation about molecular shapes.

5. Theories and laws are different kinds of scientific knowledge.

    a. I agree with this statement.
    b. I disagree with this statement.

    Explain your answer, using an example from your investigation about molecular shapes.

6. Scientists often use models to help them understand natural phenomena. Explain what a model is and why models are important, using an example from your investigation about molecular shapes.

LAB 2

7. Scientists often look for and attempt to explain patterns in nature. Explain why patterns are important, using an example from your investigation about molecular shapes.

8. In nature, the structure of an object is often related to the function or properties of that object. Explain why this is true, using an example from your investigation about molecular shapes.

Lab Handout

Lab 3. Rate of Dissolution: Why Do the Surface Area of the Solute, the Temperature of the Solvent, and the Amount of Agitation That Occurs When the Solute and the Solvent Are Mixed Affect the Rate of Dissolution?

Introduction

A solution is a uniform mixture of two or more pure substances. The substance that is dissolved is called the solute, and the substance that does the dissolving is called the solvent. When a solid dissolves in a solvent, it is assumed that the solid dissociates into the elementary particles that make up that solid. The type of elementary particle depends on the nature of the solid. A covalent compound will dissociate into individual molecules when it is added to water, whereas an ionic compound will dissociate into positive and negative ions.

Copper(II) sulfate is an example of a substance that dissolves in water. Copper(II) sulfate is an ionic compound with the chemical formula $CuSO_4$. When it is added to water it dissociates into Cu^{2+} and SO_4^{2-} ions. Copper(II) sulfate is an important agricultural chemical. Solution of copper(II) sulfate is often sprayed on plants, including wheat, potatoes, tomatoes, grapes, and citrus fruits, to help prevent fungal diseases.

Solubility, which is defined as the amount of solute that will dissolve in a given amount of solvent at a particular temperature, depends on the nature of the solute and how it interacts with the solvent. For example, 47.93 g of copper(II) sulfate will dissolve in 100 grams of water at 70°C, but only 37.46 g of sodium chloride (NaCl) will dissolve in the same amount and temperature of water. The rate of dissolution, in contrast, is a measure of how fast a solute dissolves in a solvent. There are three factors that affect the rate of dissolution: (1) the surface area of the solute, (2) the temperature of the solvent, and (3) the amount of agitation that occurs when the solute and the solvent are mixed.

To create large amounts of solutions in short periods of time, it is important to understand not only how much of a solute will dissolve in a solvent at a given temperature but also why different factors affect how fast it will dissolve. You will therefore explore three factors that affect the rate of dissolution of copper(II) sulfate and then develop a conceptual model that you can use to explain your observations and predict the dissolution rates of other solutes under different conditions.

LAB 3

Your Task

Determine how the surface area of the solute, the temperature of the solvent, and the amount of agitation that occurs when the solute and the solvent are mixed affect the rate that copper(II) sulfate dissolves in water. Then develop a conceptual model that can be used to explain *why* these factors influence the rate of dissolution. Once you have developed your conceptual model, you will need to test it to determine if it allows you to predict the dissolution rate of another solute under various conditions.

The guiding question of this investigation is, **Why do the surface area of the solute, the temperature of the solvent, and the amount of agitation that occurs when the solute and the solvent are mixed affect the rate of dissolution?**

Materials

You may use any of the following materials during your investigation:

Consumables	Equipment
• $CuSO_4$—powder • $CuSO_4$—fine crystal • $CuSO_4$—medium crystal • Rock candy	• Stopwatch • Hot plate • Electronic or triple beam balance • Stirring rod or magnetic stirrer • 1 Graduated cylinder (50 ml) • 2 Beakers (each 250 ml) • 4 Beakers or Erlenmeyer flasks (each 50 or 100 ml) • Thermometer or temperature probe • Spatula or chemical scoop • Weighing paper or dishes • Mortar and pestle

Safety Precautions

Follow all normal lab safety rules. Copper(II) sulfate is a skin and respiratory irritant and is moderately toxic by ingestion and inhalation. Your teacher will provide important information about working with the chemicals associated with this investigation. In addition, take the following safety precautions:

- Wear indirectly vented chemical-splash goggles and chemical-resistant gloves and apron while in the laboratory.
- Never taste any of the chemicals (including the rock candy).
- Handle all glassware with care.
- Use caution when working with hot plates because they can burn skin. Hot plates also need to be kept away from water and other liquids.
- Wash your hands with soap and water before leaving the laboratory.

Rate of Dissolution

Why Do the Surface Area of the Solute, the Temperature of the Solvent, and the Amount of Agitation That Occurs When the Solute and the Solvent Are Mixed Affect the Rate of Dissolution?

Investigation Proposal Required? ☐ Yes   ☐ No

Getting Started

The first step in developing your model is to design and carry out a series of experiments to determine how the surface area of the solute, the temperature of the solvent, and the amount of agitation that occurs when the solute and solvent are mixed affect the rate of dissolution of copper(II) sulfate. To conduct these experiments, you must determine what type of data you will need to collect, how you will collect the data, and how you will analyze the data to answer the guiding question.

To determine *what type of data you need to collect*, think about the following questions:

- What type of measurements or observations will you need to record during each experiment?
- When will you need to make these measurements or observations?

To determine *how you will collect the data*, think about the following questions:

- What will serve as your independent variable?
- How will you vary the independent variable while holding other variables constant?
- What types of comparisons will you need to make?
- What will you do to reduce measurement error?
- How will you keep track of the data you collect and how will you organize it?

To determine *how you will analyze the data*, think about the following questions:

- What type of calculations will you need to make?
- What type of graph could you create to help make sense of your data?

Once you have carried out your series of experiments, your group will need to develop a conceptual model to explain why these three factors influence the rate of dissolution in the way that they do. The model also needs to be able to explain the nature of the interactions that are taking place between the solute and the solvent on the submicroscopic level.

The last step in this investigation is to test your model. To accomplish this goal, you can use rock candy to determine if your model leads to accurate predictions about the rates of dissolution for a covalent compound under different conditions. If you can use your model to make accurate predictions about the rate of dissolution of other types of solutes under different conditions, then you will be able to generate the evidence you need to convince others that the conceptual model you developed is valid.

LAB 3

Connections to Crosscutting Concepts, the Nature of Science, and the Nature of Scientific Inquiry

As you work through your investigation, be sure to think about

- the importance of developing causal explanations for observations,
- how models are used to help understand natural phenomena,
- the role of imagination and creativity in science, and
- the role of experiments in science.

FIGURE L3.1
Argument presentation on a whiteboard

The Guiding Question:	
Our Claim:	
Our Evidence:	Our Justification of the Evidence:

Initial Argument

Once your group has finished collecting and analyzing your data, you will need to develop an initial argument. Your argument must include a *claim*, which is your answer to the guiding question. Your argument must also include *evidence* in support of your claim. The evidence is your analysis of the data and your interpretation of what the analysis means. Finally, you must include a *justification* of the evidence in your argument. You will therefore need to use a scientific concept or principle to explain why the evidence that you decided to use is relevant and important. You will create your initial argument on a whiteboard. Your whiteboard must include all the information shown in Figure L3.1.

Argumentation Session

The argumentation session allows all of the groups to share their arguments. One member of each group stays at the lab station to share that group's argument, while the other members of the group go to the other lab stations one at a time to listen to and critique the arguments developed by their classmates. The goal of the argumentation session is not to convince others that your argument is the best one; rather, the goal is to identify errors or instances of faulty reasoning in the initial arguments so these mistakes can be fixed. You will therefore need to evaluate the content of the claim, the quality of the evidence used to support the claim, and the strength of the justification of the evidence included in each argument that you see. To critique an argument, you might need more information than what is included on the whiteboard. You might therefore need to ask the presenter one or more follow-up questions, such as:

- How did your group collect the data? Why did you use that method?
- What did your group do to make sure the data you collected are reliable? What did you do to decrease measurement error?

Rate of Dissolution

Why Do the Surface Area of the Solute, the Temperature of the Solvent, and the Amount of Agitation That Occurs When the Solute and the Solvent Are Mixed Affect the Rate of Dissolution?

- What did your group do to analyze the data, and why did you decide to do it that way? Did you check your calculations?
- Is that the only way to interpret the results of your group's analysis? How do you know that your interpretation of the analysis is appropriate?
- Why did your group decide to present your evidence in that manner?
- What other claims did your group discuss before deciding on that one? Why did you abandon those alternative ideas?
- How confident are you that your group's claim is valid? What could you do to increase your confidence?

Once the argumentation session is complete, you will have a chance to meet with your group and revise your initial argument. Your group might need to gather more data or design a way to test one or more alternative claims as part of this process. Remember, your goal at this stage of the investigation is to develop the most valid or acceptable answer to the research question!

Report

Once you have completed your research, you will need to prepare an *investigation report* that consists of three sections that provide answers to the following questions:

1. What question were you trying to answer and why?
2. What did you do during your investigation and why did you conduct your investigation in this way?
3. What is your argument?

Your report should answer these questions in two pages or less. The report must be typed and any diagrams, figures, or tables should be embedded into the document. Be sure to write in a persuasive style; you are trying to convince others that your claim is acceptable or valid!

Checkout Questions

Lab 3. Rate of Dissolution: Why Do the Surface Area of the Solute, the Temperature of the Solvent, and the Amount of Agitation That Occurs When the Solute and the Solvent Are Mixed Affect the Rate of Dissolution?

Hikers often need more drinking water than they can carry when go on overnight trips. They can get water from rivers and streams as they hike but must purify it before they can drink it because most rivers and streams in the United States contain microorganisms that can cause serious illness. One of the safest and least expensive ways to purify water is to add an iodine tablet to it. Iodine purifies river and stream water because it kills the microorganisms in the water. The iodine tablet, however, must completely dissolve in the water before it is safe to drink.

1. What happens to a solute, such as a tablet of iodine, when it dissolves in water?

2. Use what you know about the factors that affect the rate of dissolution of a solute to suggest three things a thirsty backpacker could do to get an iodine tablet to dissolve faster in water. Be sure to explain why these three things will make the iodine tablet dissolve faster.

Rate of Dissolution

Why Do the Surface Area of the Solute, the Temperature of the Solvent, and the Amount of Agitation That Occurs When the Solute and the Solvent Are Mixed Affect the Rate of Dissolution?

3. Measuring the temperature of water is an example of an experiment.

   a. I agree with this statement.
   b. I disagree with this statement.

   Explain your answer, using an example from your investigation about rate of dissolution.

4. Scientists do not need to be creative or have a good imagination to excel in science.

   a. I agree with this statement.
   b. I disagree with this statement.

   Explain your answer, using an example from your investigation about rate of dissolution.

5. An important goal in science is to develop causal explanations for observations. Explain what a causal explanation is and why it is important, using an example from your investigation about rate of dissolution.

6. Scientists often use models to help them understand natural phenomena. Explain what a model is and why models are important, using an example from your investigation about rate of dissolution.

LAB 4

Lab Handout

Lab 4. Molarity: What Is the Mathematical Relationship Between the Moles of a Solute, the Volume of the Solvent, and the Molarity of an Aqueous Solution?

Introduction

Most of the matter around us is a mixture of pure substances. The main characteristic of a mixture is its variable composition. For example, a sports drink is a mixture of many substances, such as sugar and salt, with the proportions of substances varying depending on the type of sports drink. Mixtures can be classified as either homogeneous or heterogeneous. Homogeneous mixtures have parts that are not visually distinguishable, whereas heterogeneous mixtures have parts that can be distinguished visually. A homogeneous mixture is often called a solution. A sports drink therefore is a solution.

Much of the chemistry that affects us occurs among substances dissolved in water. It is therefore important to understand the nature of solutions in which water is the dissolving medium or the solvent. This type of solution is called an aqueous solution. An aqueous solution contains one or more chemicals (or *solutes*) dissolved in water (the *solvent*). The most common way to describe the concentration of a solute in an aqueous solution is to use a unit of measurement called *molarity*. In this lab investigation, you will explore the relationship between moles of solute, volume of solvent, and molarity.

Your Task

Use a computer simulation to determine the mathematical relationship between moles of solute, volume of solvent, and molarity. Once you have determined this relationship, you should be able to set up various functions that will allow you to accurately predict

- the molarity of a solution given the moles of solute and solvent volume,
- the moles of solute given the molarity of the solution and the volume of the solvent, and
- the volume of the solvent given the molarity of the solution and the moles of the solute.

The guiding question of this investigation is, **What is the mathematical relationship between the moles of a solute, the volume of the solvent, and the molarity of an aqueous solution?**

Molarity

What Is the Mathematical Relationship Between the Moles of a Solute, the Volume of the Solvent, and the Molarity of an Aqueous Solution?

Materials

You will use an online simulation called *Molarity* to conduct your investigation. You can access the simulation by going to the following website: *http://phet.colorado.edu/en/simulation/molarity*.

Safety Precautions

Follow all normal lab safety rules.

Investigation Proposal Required?   ☐ Yes   ☐ No

Getting Started

The first step in developing your mathematical function is to determine how moles of solute and volume of the solvent are related to the molarity of a solution. The *Molarity* simulation (see screenshot in Figure L4.1) allows you to mix different moles of solute in different volumes of water (the solvent). It then provides a measure of the molarity of the resulting aqueous solution.

FIGURE L4.1
A screenshot from the *Molarity* simulation

Before you start using the simulation, you must determine what type of data you will need to collect, how you will collect the data, and how you will analyze the data to answer the guiding question.

To determine *what type of data you need to collect*, think about the following questions:

- What type of observations will you need to record during your investigation?

- When will you need to make these observations?

To determine *how you will collect the data*, think about the following questions:

- What types of comparisons will you need to make?
- How will you keep track of the data you collect and how will you organize it?

To determine *how you will analyze the data*, think about the following questions:

- What type of calculations will you need to make?
- What type of graph could you create to help make sense of your data?

Once you have collected and analyzed your data, your group will need to develop a function that can be used to predict (1) the molarity of a solution given the moles of solute and solvent volume, (2) the moles of solute given the molarity of the solution and the volume of the solvent, and (3) the volume of the solvent given the molarity of the solution and the moles of the solute. You will then need to test your function using the simulation. If you are able to use your function to make accurate predictions, then you will be able to generate the evidence you need to convince others that the function you developed is valid.

Connections to Crosscutting Concepts, the Nature of Science, and the Nature of Scientific Inquiry

As you work through your investigation, be sure to think about

- the importance of looking for proportional relationships between different quantities,
- why it is important to track what happens to matter within a system,
- the difference between data and evidence in science, and
- the wide range of methods that can be used during a scientific investigation.

FIGURE L4.2

Argument presentation on a whiteboard

The Guiding Question:	
Our Claim:	
Our Evidence:	Our Justification of the Evidence:

Initial Argument

Once your group has finished collecting and analyzing your data, you will need to develop an initial argument. Your argument must include a *claim*, which is your answer to the guiding question. Your argument must also include *evidence* in support of your claim. The evidence is your analysis of the data and your interpretation of what the analysis means. Finally, you must include a *justification* of the evidence in your argument. You will therefore need to use a scientific concept or principle to explain why the evidence that you decided to use is relevant and important. You will create your initial argument on a whiteboard. Your whiteboard must include all the information shown in Figure L4.2.

Molarity

What Is the Mathematical Relationship Between the Moles of a Solute, the Volume of the Solvent, and the Molarity of an Aqueous Solution?

Argumentation Session

The argumentation session allows all of the groups to share their arguments. One member of each group stays at the lab station to share that group's argument, while the other members of the group go to the other lab stations one at a time to listen to and critique the arguments developed by their classmates. The goal of the argumentation session is not to convince others that your argument is the best one; rather, the goal is to identify errors or instances of faulty reasoning in the initial arguments so these mistakes can be fixed. You will therefore need to evaluate the content of the claim, the quality of the evidence used to support the claim, and the strength of the justification of the evidence included in each argument that you see. To critique an argument, you might need more information than what is included on the whiteboard. You might therefore need to ask the presenter one or more follow-up questions, such as:

- What did your group do to analyze the data, and why did you decide to do it that way?
- Is that the only way to interpret the results of your group's analysis? How do you know that your interpretation of the analysis is appropriate?
- Why did your group decide to present your evidence in that manner?
- What other claims did your group discuss before deciding on that one? Why did you abandon those alternative ideas?
- How confident are you that your group's claim is valid? What could you do to increase your confidence?

Once the argumentation session is complete, you will have a chance to meet with your group and revise your original argument. Your group might need to gather more data or design a way to test one or more alternative claims as part of this process. Remember, your goal at this stage of the investigation is to develop the most valid or acceptable answer to the research question!

Report

Once you have completed your research, you will need to prepare an *investigation report* that consists of three sections that provide answers to the following questions:

1. What question were you trying to answer and why?
2. What did you do during your investigation and why did you conduct your investigation in this way?
3. What is your argument?

LAB 4

Your report should answer these questions in two pages or less. The report must be typed and any diagrams, figures, or tables should be embedded into the document. Be sure to write in a persuasive style; you are trying to convince others that your claim is acceptable or valid!

Checkout Questions

Lab 4. Molarity: What Is the Mathematical Relationship Between the Moles of a Solute, the Volume of the Solvent, and the Molarity of an Aqueous Solution?

1. Describe the mathematical relationship between the moles of a solute, the volume of a solvent, and the molarity of an aqueous solution.

2. Some household cleaners come in concentrations stronger than necessary for basic cleaning jobs. Jeremy followed the instructions on a cleaning bottle and mixed enough cleaner with 4.0 L of water to form a 1.0 M cleaning solution. After testing his cleaning solution, he decided he should double the concentration for a tough stain. Jeremy added the same amount of cleaner and then another 4.0 L of water to his bucket.

    Using what you know about molarity, explain why Jeremy did not succeed in doubling the concentration of his cleaning solution.

3. All scientific investigations are experiments.
    a. I agree with this statement.
    b. I disagree with this statement.

    Explain your answer, using an example from your investigation about molarity.

LAB 4

4. Evidence is data that support a claim.

    a. I agree with this statement.
    b. I disagree with this statement.

    Explain your answer, using an example from your investigation about molarity.

5. Scientists often need to look for proportional relationships between different quantities during an investigation. Explain what a proportional relationship is and why it is important, using an example from your investigation about molarity.

6. It is often important to track how matter flows into, out of, and within a system during an investigation. Explain why it is important to keep track of matter when studying a system, using an example from your investigation about molarity.

Lab Handout

Lab 5. Temperature Changes Due to Evaporation: Which of the Available Substances Has the Strongest Intermolecular Forces?

Introduction

Matter exists in three basic states: solid, liquid, and gas. Whether a substance is a solid, liquid, or gas at room temperature (20°C–25°C) depends on the properties of that specific substance. Oxygen, for instance, is a gas at room temperature, but water is a liquid. Substances also maintain their state over a broad range of conditions; for example, water is a liquid from 0°C all the way to 100°C. For a substance to transition from one phase to another, the conditions again must be just right. Transitioning from a solid to a liquid (i.e., melting) or transitioning from a liquid to a gas (i.e., boiling) requires a transfer of energy.

Boiling or vaporization is the process by which a substance changes from a liquid to a gas. Evaporation is vaporization that occurs at the surface of a liquid. Chemists explain the process of evaporation using the *molecular-kinetic theory of matter*. According to this theory, the molecules that make up a liquid are in constant motion but are attracted to other molecules by different types of *intermolecular forces*. The temperature of the liquid is proportional to the average *kinetic energy* of the molecules found within that liquid. Kinetic energy is the energy of motion. The kinetic energy of a molecule depends on its velocity and its mass. From this perspective, some of the molecules at the surface of a liquid will have greater or lower kinetic energy than the average. Some of the molecules will have enough kinetic energy to disrupt the intermolecular forces that hold it near the other molecules in the liquid. These molecules, as a result, will break away from the surface of the liquid. When this happens, the average kinetic energy of the remaining molecules in the liquid decreases. The temperature of a liquid therefore goes down as it evaporates.

There are several important factors that will influence the evaporation rate of a liquid. One of these factors is temperature. A liquid will evaporate faster at a higher temperature because more molecules at the surface will have enough kinetic energy to break free from the other molecules in that liquid. Another important factor is the type of intermolecular forces that exist between the molecules in that liquid. Intermolecular forces, as noted earlier, are attractive in nature. A liquid with weak intermolecular forces will evaporate quickly because it takes less kinetic energy for a molecule at the surface of the liquid to break away from the other molecules in the liquid. A liquid with strong intermolecular forces, in contrast, will evaporate slowly because the molecules that make up this type of

LAB 5

liquid require more kinetic energy to break away from the other molecules at the surface of the liquid.

In this investigation, you will use what you have learned about evaporation, the molecular-kinetic theory of matter, and this brief introduction to intermolecular forces to determine the relative strength of the intermolecular forces that exist between the molecules in six different types of liquids. This is important for you to be able to do because chemists often need to rely on their understanding of intermolecular interactions to explain the bulk properties of matter. It is also important because it will help you understand the underlying reasons for several important natural phenomena, such as the water cycle or evaporative cooling, that have many practical applications for everyday life.

Your Task

Determine how temperature changes when six different liquids evaporate from a material and then rank the substances in terms of the relative strength of their intermolecular forces.

The guiding question of this investigation is, **Which of the available substances has the strongest intermolecular forces?**

Materials

You may use any of the following materials during your investigation:

Consumables	Equipment
• Filter paper • Rubber bands • Acetone, $(CH_3)_2CO$ • Ethyl alcohol, CH_3CH_2OH • Heptane, C_7H_{16} • Hexane, C_6H_{14} • Isopropyl alcohol, $(CH3)_2CHOH$ • Methyl alcohol, CH_3OH	• Temperature sensor kit • Test tubes w/ stoppers • Test tube rack • Pipettes

Safety Precautions

Follow all normal lab safety rules. Acetone, ethyl alcohol, heptane, hexane, isopropyl alcohol, and methyl alcohol are flammable liquids. Acetone, ethyl alcohol, and heptane are each moderately toxic by ingestion and inhalation. Given the fire and health hazards associated with these chemicals, all lab work must be done under a fume hood. In addition, take the following safety precautions:

- Wear indirectly vented chemical-splash goggles and chemical-resistant gloves and apron while in the laboratory.
- Handle all glassware with care.
- Wash your hands with soap and water before leaving the laboratory.

Temperature Changes Due to Evaporation
Which of the Available Substances Has the Strongest Intermolecular Forces?

Investigation Proposal Required?   ☐ Yes   ☐ No

Getting Started

The first step in this investigation is to carry out a series of tests to determine how the temperature of each substance changes as it evaporates. One way to generate these data is by attaching a small piece of filter paper to a temperature probe using a rubber band (see Figure L5.1). Then briefly soak the filter paper in one of the substances, remove the apparatus from the liquid, and observe as the liquid evaporates from the filter paper.

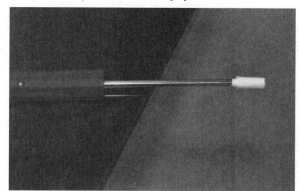

FIGURE L5.1
Temperature probe with filter paper

Before you begin your tests, however, you will need to determine what type of data you need to collect during each test, how you will collect the data, and how you will analyze the data.

To determine *what type of data you need to collect*, think about the following questions:

- What type of measurements or observations will you need to record during each test?
- How often will you need to make these measurements or observations?

To determine *how you will collect the data*, think about the following questions:

- How long will you need to conduct each test?
- What types of comparisons will you need to make?
- What will you do to reduce measurement error?
- How will you keep track of the data you collect and how will you organize it?

To determine *how you will analyze the data*, think about the following questions:

- What type of calculations will you need to make?
- What type of graph could you create to help make sense of your data?

The second step in your investigation will be to develop a way to rank order the substances in terms of the relative strength of their intermolecular forces. At this point in the investigation, it may also be useful to consider the molecular characteristics of each substance, such as the molecular structure or shape, molar mass, whether the molecule is polar, and so on. Table L5.1 provides the structural formula for each substance for comparison purposes.

Student Lab Manual for Argument-Driven Inquiry in Chemistry: Lab Investigations for Grades 9–12

LAB 5

TABLE L5.1
Molar mass and the structural formulas of the substances used in this investigation

Substance	Molar mass	Structural formula
Acetone	58.08 g/mol	H₃C—C(=O)—CH₃
Ethyl alcohol	46.07 g/mol	CH₃—CH₂—OH
Heptane	100.21 g/mol	CH₃—CH₂—CH₂—CH₂—CH₂—CH₂—CH₃
Hexane	86.18 g/mol	CH₃—CH₂—CH₂—CH₂—CH₂—CH₃
Isopropyl alcohol	60.10 g/mol	CH₃—CH(OH)—CH₃
Methyl alcohol	32.04 g/mol	CH₃—OH

Temperature Changes Due to Evaporation
Which of the Available Substances Has the Strongest Intermolecular Forces?

Connections to Crosscutting Concepts, the Nature of Science, and the Nature of Scientific Inquiry

As you work through your investigation, be sure to think about

- how energy and matter flow within systems,
- the importance of the structure of molecules in relation to the ways they function,
- the nature and role of experiments in science, and
- the difference between observations and inferences.

Initial Argument

Once your group has finished collecting and analyzing your data, you will need to develop an initial argument. Your argument must include a *claim*, which is your answer to the guiding question. Your argument must also include *evidence* in support of your claim. The evidence is your analysis of the data and your interpretation of what the analysis means. Finally, you must include a *justification* of the evidence in your argument. You will therefore need to use a scientific concept or principle to explain why the evidence that you decided to use is relevant and important. You will create your initial argument on a whiteboard. Your whiteboard must include all the information shown in Figure L5.2.

FIGURE L5.2
Argument presentation on a whiteboard

The Guiding Question:	
Our Claim:	
Our Evidence:	Our Justification of the Evidence:

Argumentation Session

The argumentation session allows all of the groups to share their arguments. One member of each group stays at the lab station to share that group's argument, while the other members of the group go to the other lab stations one at a time to listen to and critique the arguments developed by their classmates. The goal of the argumentation session is not to convince others that your argument is the best one; rather, the goal is to identify errors or instances of faulty reasoning in the initial arguments so these mistakes can be fixed. You will therefore need to evaluate the content of the claim, the quality of the evidence used to support the claim, and the strength of the justification of the evidence included in each argument that you see. To critique an argument, you might need more information than what is included on the whiteboard. You might therefore need to ask the presenter one or more follow-up questions, such as:

- What did your group do to analyze the data, and why did you decide to do it that way?
- Is that the only way to interpret the results of your group's analysis? How do you know that your interpretation of the analysis is appropriate?

LAB 5

- Why did your group decide to present your evidence in that manner?
- What other claims did your group discuss before deciding on that one? Why did you abandon those alternative ideas?
- How confident are you that your group's claim is valid? What could you do to increase your confidence?

Once the argumentation session is complete, you will have a chance to meet with your group and revise your original argument. Your group might need to gather more data or design a way to test one or more alternative claims as part of this process. Remember, your goal at this stage of the investigation is to develop the most valid or acceptable answer to the research question!

Report

Once you have completed your research, you will need to prepare an *investigation report* that consists of three sections that provide answers to the following questions:

1. What question were you trying to answer and why?
2. What did you do during your investigation and why did you conduct your investigation in this way?
3. What is your argument?

Your report should answer these questions in two pages or less. The report must be typed and any diagrams, figures, or tables should be embedded into the document. Be sure to write in a persuasive style; you are trying to convince others that your claim is acceptable or valid!

Checkout Questions

Lab 5. Temperature Changes Due to Evaporation: Which of the Available Substances Has the Strongest Intermolecular Forces?

1. Describe how the strength of intermolecular forces is related to the boiling point of a liquid.

2. Our bodies generate sweat to help regulate our temperature and stay cool on hot days or during exercise. Many people think that the sweat is cool and in turn cools our bodies, but the sweat our bodies produce is actually the same temperature as the inside of our body where the sweat was generated. Sweat actually cools our skin due to evaporation.

    Use what you know about the process of evaporation to explain how evaporative cooling works to keep our skin cool.

3. If two scientists observe the same event, it is likely that they will come to the same conclusions.

    a. I agree with this statement.
    b. I disagree with this statement.

    Explain your answer, using an example from your investigation about temperature changes due to evaporation.

LAB 5

4. Conducting an experiment is one way to investigate questions in science, but there are other ways to conduct scientific investigations.

    a. I agree with this statement.
    b. I disagree with this statement.

    Explain your answer, using an example from your investigation about temperature changes due to evaporation.

5. Understanding how matter and energy flow within and between systems is important in science. Explain why this is important, using an example from your investigation about temperature changes due to evaporation.

6. Some scientists devote their career to understanding the structure and function of just a handful of molecules. Explain why understanding the structure and function of a molecule is important, using an example from your investigation about temperature changes due to evaporation.

Lab Handout

Lab 6. Pressure, Temperature, and Volume of Gases: How Does Changing the Volume or Temperature of a Gas Affect the Pressure of That Gas?

Introduction

There are three states of matter: solid, liquid, and gas. Each state of matter has physical properties that distinguish it from the other states; for example, matter in the solid phase has a definite shape, whereas matter in the liquid or gas phase will take on the shape of its container. The physical properties associated with the states of matter allow us to predict how different substances may react under various conditions. Particles in a gas move about more freely than those in a solid or liquid and therefore react to changes in temperature and pressure in a manner that is different than solids or liquids.

The *volume* of a sample of gas, or the amount of space that a sample of gas occupies, is particularly influenced by a variety of factors such as temperature or pressure. Just like the shape of a sample of gas or liquid is determined by its container, the volume of a sample of gas is influenced by its surroundings. A small sample of gas, like air, may be confined to a small container such as a balloon, or if the balloon pops the sample of gas can expand to occupy the entire volume of a classroom. Consider a tank of helium gas used to fill birthday balloons. There is a large amount of gas stored inside the tank, but several birthday balloons filled with a sample of the gas can easily expand to a size much larger than the tank. Understanding the physical properties of gases and how a gas interacts with its surroundings helps to explain this phenomenon. In this investigation you will explore the relationship between volume, temperature, and pressure for a gas within a closed system.

Your Task

Determine how changes to the volume and the temperature of a gas within a closed system affect the pressure of that gas. Then develop a general mathematical model that can be used to apply and describe these relationships with respect to all gases.

The guiding question for this lab is, **How does changing the volume or temperature of a gas affect the pressure of that gas?**

LAB 6

Materials

You may use any of the following materials during your investigation:

Consumable	Equipment
• Ice	• Gas pressure sensor • Temperature sensor • Sensor interface • Syringe • Erlenmeyer flask • Single-hole rubber stopper • Rubber tubing • Beaker (500 ml) • Hot plate

Safety Precautions

Follow all normal lab safety rules. In addition, take the following safety precautions:

- Wear indirectly vented chemical-splash goggles and chemical-resistant gloves and apron while in the laboratory.
- Handle all glassware with care.
- Use caution when working with hot plates because they can burn skin. Hot plates also need to be kept away from water and other liquids.
- Wash your hands with soap and water before leaving the laboratory.

Investigation Proposal Required?   ☐ Yes   ☐ No

Getting Started

To determine the relationship between the pressure, the volume, and the temperature of a gas, you will need to set up an apparatus that will allow you to first measure changes in gas pressure when the volume of gas changes. This can be accomplished with the apparatus shown in Figure L6.1 (p. 54). You will then need to be able to measure changes in gas pressure when the temperature of the gas changes. This can be accomplished with the apparatus shown in Figure L6.2 (p. 54). Once you have set up these apparatuses, you must determine what type of data you need to collect, how you will collect the data, and how you will analyze the data.

To determine *what type of data you need to collect*, think about the following questions:

- What type of measurements or observations will you need to record during your investigation?
- When will you need to make these measurements or observations?

To determine *how you will collect the data*, think about the following questions:

Pressure, Temperature, and Volume of Gases
How Does Changing the Volume or Temperature of a Gas Affect the Pressure of That Gas?

- What will serve as your dependent variable(s)?
- What will serve as a control (or comparison) condition?
- What types of treatment conditions will you need to set up and how will you do it?
- How will you make sure that your data are of high quality (i.e., how will you reduce error)?

To determine *how you will analyze the data*, think about the following questions:

- How will you determine if there is a difference between the treatment conditions and the control condition?
- What type of calculations will you need to make?
- What type of graph could you create to help make sense of your data?

Once you have finished collecting your data, your group will need to develop a mathematical model that describes how the pressure of a gas is affected by changes in the volume and the temperature. When developing a mathematical model, variables that are inversely related are multiplied and variables that are directly related are divided. Keep these mathematical relationships in mind as you develop your model.

The last step in this investigation is to test your model. To accomplish this goal, you can use your model to make predictions about the pressure of a gas in a closed system under different conditions. If you are able to make accurate predictions with your model, then you will be able to generate the evidence you need to convince others that your model is valid.

Connections to Crosscutting Concepts, the Nature of Science, and the Nature of Scientific Inquiry

As you work through your investigation, be sure to think about

- the importance of developing causal explanations for observations,
- how models are used to help understand natural phenomena,
- the difference between laws and theories in science, and
- the difference between data and evidence in science.

Initial Argument

Once your group has finished collecting and analyzing your data, you will need to develop an initial argument. Your argument must include a *claim*, which is your answer to the guiding question. Your argument must also include *evidence* in support of your claim. The evidence is your analysis of the data and your interpretation of what the analysis means. Finally, you must include a *justification* of the evidence in your argument. You will therefore need to use a scientific concept or principle to explain why the evidence that

LAB 6

FIGURE L6.1
Apparatus used to measure changes in gas pressure in response to changes in the volume of the gas

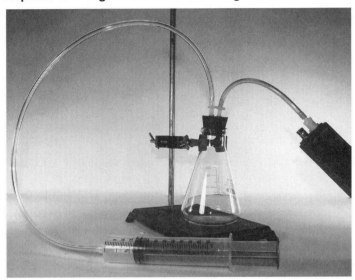

FIGURE L6.2
Apparatus used to measure changes in gas pressure in response to changes in the temperature of the gas

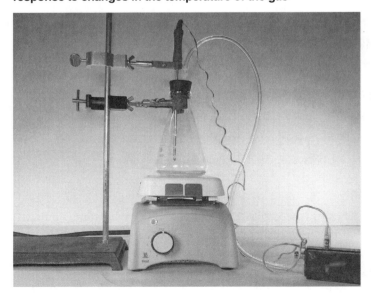

Pressure, Temperature, and Volume of Gases
How Does Changing the Volume or Temperature of a Gas Affect the Pressure of That Gas?

you decided to use is relevant and important. You will create your initial argument on a whiteboard. Your whiteboard must include all the information shown in Figure L6.3.

FIGURE L6.3
Argument presentation on a whiteboard

The Guiding Question:	
Our Claim:	
Our Evidence:	Our Justification of the Evidence:

Argumentation Session

The argumentation session allows all of the groups to share their arguments. One member of each group stays at the lab station to share that group's argument, while the other members of the group go to the other lab stations one at a time to listen to and critique the arguments developed by their classmates. The goal of the argumentation session is not to convince others that your argument is the best one; rather, the goal is to identify errors or instances of faulty reasoning in the initial arguments so these mistakes can be fixed. You will therefore need to evaluate the content of the claim, the quality of the evidence used to support the claim, and the strength of the justification of the evidence included in each argument that you see. To critique an argument, you might need more information than what is included on the whiteboard. You might therefore need to ask the presenter one or more follow-up questions, such as:

- What did your group do to analyze the data, and why did you decide to do it that way?
- Is that the only way to interpret the results of your group's analysis? How do you know that your interpretation of the analysis is appropriate?
- Why did your group decide to present your evidence in that manner?
- What other claims did your group discuss before deciding on that one? Why did you abandon those alternative ideas?
- How confident are you that your group's claim is valid? What could you do to increase your confidence?

Once the argumentation session is complete, you will have a chance to meet with your group and revise your original argument. Your group might need to gather more data or design a way to test one or more alternative claims as part of this process. Remember, your goal at this stage of the investigation is to develop the most valid or acceptable answer to the research question!

Report

Once you have completed your research, you will need to prepare an *investigation report* that consists of three sections that provide answers to the following questions:

1. What question were you trying to answer and why?

LAB 6

2. What did you do during your investigation and why did you conduct your investigation in this way?
3. What is your argument?

Your report should answer these questions in two pages or less. The report must be typed and any diagrams, figures, or tables should be embedded into the document. Be sure to write in a persuasive style; you are trying to convince others that your claim is acceptable or valid!

Checkout Questions

Lab 6. Pressure, Temperature, and Volume of Gases: How Does Changing the Volume or Temperature of a Gas Affect the Pressure of That Gas?

1. Describe the relationship between pressure and volume for a gas and volume and temperature for a gas. What are the mathematical equations that describe these relationships?

2. Susan releases a birthday balloon filled with helium into the air. The balloon quickly rises into the sky until it is out of sight. Gradually as the balloon rises, it begins to expand. Eventually, the balloon gets so large that it pops.

    Use what you know about the relationships between the pressure, volume, and temperature of a gas to explain what conditions must have been present for the balloon to expand.

3. Data and evidence are the same thing in science.

    a. I agree with this statement.
    b. I disagree with this statement.

    Explain your answer, using an example from your investigation about pressure, temperature, and volume of gases.

LAB 6

4. It is common for a theory to become a law in science.

   a. I agree with this statement.
   b. I disagree with this statement.

   Explain your answer, using an example from your investigation about pressure, temperature, and volume of gases.

5. Scientists often generate causal explanations for the observations they make. Explain why this is important, using an example from your investigation about pressure, temperature, and volume of gases.

6. Scientists create models to help understand natural phenomena. Explain what a model is and why models are useful, using an example from your investigation about pressure, temperature, and volume of gases.

Lab Handout

Lab 7. Periodic Trends: Which Properties of the Elements Follow a Periodic Trend?

Introduction

Periodic trends are the tendencies of certain properties of the elements to increase or decrease as you progress along a row or a column of the periodic table. A row in the periodic table is called a *period*, and a column in the periodic table is called a *group*. These trends can occur in both physical and atomic properties of the elements. The periodic table is organized in a way that makes these trends relatively easy to determine. Scientists can use these trends to help them predict an element's properties, which can determine how it will react in certain situations. The similar atomic structure among elements in a group or period helps explain how these trends occur.

Atomic radius is an example of a property that has a periodic trend; as illustrated in Figure L7.1, atomic radius increases as you move down the periodic table regardless of group. Some properties, however, only change in a uniform manner within a specific group. These properties are often described as having a *quasi-periodic trend*. Boiling point is an example of a property that has a quasi-periodic trend; boiling point only changes in a uniform manner within a group, but it does not follow a similar pattern when you look across a period. In this investigation, you will explore how the physical and atomic properties of the element change across periods and groups in order to identify the other periodic trends.

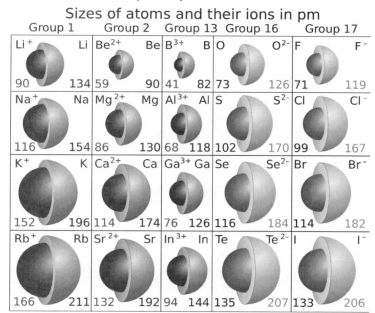

FIGURE L7.1

Atomic radius is an example of a periodic trend.

Your Task

You will be given an Excel file that includes a list of the known elements and information about the atomic mass, density, melting point, specific heat capacity, and electronegativity

LAB 7

of the elements. You must use this information along with the definitions provided in the "Introduction" to determine which of these properties follow a periodic trend and which ones do not.

The guiding question of this investigation is, **Which properties of the elements follow a periodic trend?**

Materials
You may use any of the following materials during your investigation:

- Computer
- "Properties of Elements" Excel file
- Periodic table

Safety Precautions
Follow all normal lab safety rules.

Investigation Proposal Required?   ☐ Yes   ☐ No

Getting Started
To answer the guiding question, you will need to analyze an existing data set. To accomplish this task, you must determine what type of data you need to examine and how you will analyze the data.

To determine *what type of data you will need to examine*, think about the following questions:

- What makes something a periodic trend or a quasi-period trend?
- Which elements will you need to look up to establish a trend?

To determine *how you will analyze the data*, think about the following questions:

- What type of graph could you create to help make sense of your data?
- What types of calculations will you need to make?

Connections to Crosscutting Concepts, the Nature of Science, and the Nature of Scientific Inquiry
As you work through your investigation, be sure to think about

- the importance of identifying patterns in science,
- the importance of proportional relationships,

Periodic Trends
Which Properties of the Elements Follow a Periodic Trend?

- the difference between observations and inferences, and
- how science is influenced by social and cultural values.

Initial Argument

Once your group has finished collecting and analyzing your data, you will need to develop an initial argument. Your argument must include a *claim*, which is your answer to the guiding question. Your argument must also include *evidence* in support of your claim. The evidence is your analysis of the data and your interpretation of what the analysis means. Finally, you must include a *justification* of the evidence in your argument. You will therefore need to use a scientific concept or principle to explain why the evidence that you decided to use is relevant and important. You will create your initial argument on a whiteboard. Your whiteboard must include all the information shown in Figure L7.2.

FIGURE L7.2
Argument presentation on a whiteboard

The Guiding Question:	
Our Claim:	
Our Evidence:	Our Justification of the Evidence:

Argumentation Session

The argumentation session allows all of the groups to share their arguments. One member of each group stays at the lab station to share that group's argument, while the other members of the group go to the other lab stations one at a time to listen to and critique the arguments developed by their classmates. The goal of the argumentation session is not to convince others that your argument is the best one; rather, the goal is to identify errors or instances of faulty reasoning in the initial arguments so these mistakes can be fixed. You will therefore need to evaluate the content of the claim, the quality of the evidence used to support the claim, and the strength of the justification of the evidence included in each argument that you see. To critique an argument, you might need more information than what is included on the whiteboard. You might therefore need to ask the presenter one or more follow-up questions, such as:

- What did your group do to analyze the data, and why did you decide to do it that way?
- Is that the only way to interpret the results of your group's analysis? How do you know that your interpretation of the analysis is appropriate?
- Why did your group decide to present your evidence in that manner?
- What other claims did your group discuss before deciding on that one? Why did you abandon those alternative ideas?
- How confident are you that your group's claim is valid? What could you do to increase your confidence?

LAB 7

Once the argumentation session is complete, you will have a chance to meet with your group and revise your original argument. Your group might need to gather more data or design a way to test one or more alternative claims as part of this process. Remember, your goal at this stage of the investigation is to develop the most valid or acceptable answer to the research question!

Report

Once you have completed your research, you will need to prepare an *investigation report* that consists of three sections that provide answers to the following questions:

1. What question were you trying to answer and why?
2. What did you do during your investigation and why did you conduct your investigation in this way?
3. What is your argument?

Your report should answer these questions in two pages or less. The report must be typed and any diagrams, figures, or tables should be embedded into the document. Be sure to write in a persuasive style; you are trying to convince others that your claim is acceptable or valid!

Checkout Questions

Lab 7. Periodic Trends: Which Properties of the Elements Follow a Periodic Trend?

1. Describe the difference between a periodic and a quasi-periodic trend.

2. Fluorine (F), boron (B), and lithium (Li) are all in the same period, but are in different groups. Fluorine has a higher electronegativity than boron, which has a higher electronegativity than lithium (F = 3.98, B = 2.04, Li = 0.98; all using the Pauling scale). However, boron has a higher melting point than both fluorine and lithium, and lithium's melting point is higher than that of fluorine (F = −220°C, B = 2300°C, Li = 180°C).

   Using your knowledge of periodic trends and the information above, describe the nature of the periodic trends for electronegativity and melting point.

3. "The element has high ionization energy" is an example of an observation.

   a. I agree with this statement.
   b. I disagree with this statement.

   Explain your answer, using an example from your investigation about periodic trends.

LAB 7

4. Cultural values and expectations affect how scientists conduct an investigation.

   a. I agree with this statement.

   b. I disagree with this statement.

   Explain your answer, using an example from your investigation about periodic trends.

5. Scientists often look for and attempt to explain patterns in nature. Explain why patterns are important, using an example from your investigation about periodic trends.

6. Scientists often need to look for proportional relationships. Explain why looking for a proportional relationship is often useful in science, using an example from your investigation about periodic trends.

Lab Handout

Lab 8. Solutes and the Freezing Point of Water: How Does the Addition of Different Types of Solutes Affect the Freezing Point of Water?

Introduction

When substances undergo a change in phase, whether it is solid to liquid, liquid to gas, or the reverse, a transfer of energy must occur. Phase changes can be described as *endothermic* (absorbing energy) or *exothermic* (releasing energy) processes. Consider what happens when water freezes. When water changes from a liquid to a solid, the molecules lose energy and become more ordered; thus, freezing is an exothermic process because energy is released. Alternatively, when ice melts or water boils, the ice or liquid water gains or absorbs energy from its surroundings; thus, melting and boiling are endothermic processes.

A graph of a cooling curve provides a visual representation of what happens during phase changes. Figure L8.1 shows a cooling curve for a sample of water being cooled at a constant rate. The sample of water starts at 120°C in the gas phase. As it is cooled, the temperature of the sample decreases because the average kinetic energy of the water molecules in the sample is decreasing over time. At 100°C, the cooling curve levels off and the temperature remains constant until all of the steam has condensed into a liquid. This period of time represents the *condensation point*. At this point, the water molecules no longer have enough kinetic energy to remain in the gas phase. The liquid water then begins to decrease in temperature again as it loses more kinetic energy. At the freezing point, 0°C, the temperature of the sample

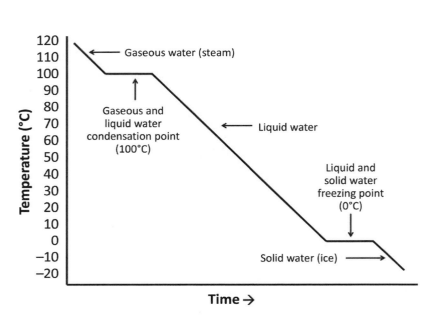

FIGURE L8.1

Cooling curve for water

LAB 8

levels off again and the liquid water transitions to a solid state. Once all the liquid water has become ice, the sample will once again decrease in temperature as it continues to lose energy.

When a solute is added to water, some of its physical properties may change. An example of a physical property of water that can change with the addition of a solute is its ability to conduct electricity. Pure water does not conduct electricity. However, when an *electrolyte* is added to water, the resulting solution is able to conduct electricity. Water, however, will not conduct electricity when a *non-electrolyte* is added to it. Electrolytes are ionic compounds such as sodium chloride and potassium chloride. Ionic compounds consist of ions joined together by ionic bonds, but these compounds dissociate into individual ions when mixed with water. Non-electrolytes, in contrast, are molecular compounds such as sucrose and propanol. Molecular compounds are made up of molecules that consist of atoms joined together by covalent bonds. The molecules found within a molecular compound do not break down when the compound is added to water; the molecules simply dissociate from each other as a result of the dissolution process. Other physical properties of water, such as its freezing point or boiling point, may also change in different ways when different types of solutes are added to it. In this investigation, you will determine how the addition of different types of solutes affects the freezing point of water.

Your Task

Determine how the addition of electrolytes and non-electrolytes affects the freezing point of water. Your group may test any of the following solutes:

Non-electrolytes	Electrolytes
• Glycerin ($C_3H_8O_3$) • 2-Propanol (C_3H_8O) • Sucrose ($C_{12}H_{22}O_{11}$)	• Potassium chloride (KCl) • Sodium chloride (NaCl) • Sodium nitrate ($NaNO_3$)

The guiding question of this investigation is, **How does the addition of different types of solutes affect the freezing point of water?**

Materials

You may use any of the following materials during your investigation:

Consumables	Equipment
• $C_3H_8O_3$ • C_3H_8O • $C_{12}H_{22}O_{11}$ • KCl • NaCl • $NaNO_3$ • Rock salt • Ice • Distilled water	• Temperature probe • Styrofoam cup • Test tube (medium: 20 × 150 mm) • Erlenmeyer flasks • Graduated cylinder (100 ml) • Parafilm

Solutes and the Freezing Point of Water
How Does the Addition of Different Types of Solutes Affect the Freezing Point of Water?

Safety Precautions

Follow all normal lab safety rules. 2-Propanol is flammable so be sure to keep it away from flames. Your teacher will explain relevant and important information about working with the chemicals associated with this investigation. In addition, take the following safety precautions:

- Wear indirectly vented chemical-splash goggles and chemical-resistant gloves and apron while in the laboratory.
- Handle all glassware with care.
- Wash your hands with soap and water before leaving the laboratory.

Investigation Proposal Required? ☐ Yes ☐ No

Getting Started

To measure the freezing point of distilled water or a solution made from one of the solutes, you can use rock salt and ice to create a supercool environment (less than 0°C). The basic setup is illustrated in Figure L8.2. The aqueous solutions that you test should have a concentration of at least 0.5 M and no more than 1.0 M. When you make your solutions, use 100 ml of H_2O and prepare them in an Erlenmeyer flask. Be sure to swirl the mixture until all the solute dissolves.

Now that you know how to measure the freezing point of distilled water or an aqueous solution using this equipment, you will need to design an investigation to answer the guiding question. You will therefore need to think about what type of data you need to collect, how you will collect the data, and how you will analyze the data.

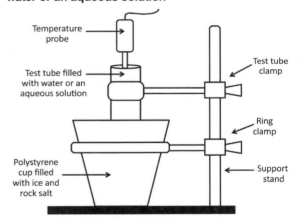

FIGURE L8.2

Equipment used to measure the freezing point of water or an aqueous solution

To determine *what type of data you need to collect*, think about the following questions:

- How will you know when an aqueous solution has reached its freezing point?
- What type of measurements will you need to record during your investigation?

To determine *how you will collect the data*, think about the following questions:

- What will serve as a control (or comparison) condition?
- What types of treatment conditions will you need to set up and how will you do it?
- What will you need to keep constant across comparisons?

LAB 8

- How will you make sure that your data are of high quality (i.e., how will you reduce error)?

To determine *how you will analyze the data*, think about the following questions:

- How will you determine if there is a difference between the treatment conditions and the control condition?
- What type of table could you create to help make sense of your data?

Connections to Crosscutting Concepts, the Nature of Science, and the Nature of Scientific Inquiry

As you work through your investigation, be sure to think about

- the importance of identifying patterns,
- how energy and matter are related to each other,
- how imagination and creativity are used during your investigation, and
- the nature and role of experiments.

FIGURE L8.3
Argument presentation on a whiteboard

The Guiding Question:	
Our Claim:	
Our Evidence:	Our Justification of the Evidence:

Initial Argument

Once your group has finished collecting and analyzing your data, you will need to develop an initial argument. Your argument must include a *claim*, which is your answer to the guiding question. Your argument must also include *evidence* in support of your claim. The evidence is your analysis of the data and your interpretation of what the analysis means. Finally, you must include a *justification* of the evidence in your argument. You will therefore need to use a scientific concept or principle to explain why the evidence that you decided to use is relevant and important. You will create your initial argument on a whiteboard. Your whiteboard must include all the information shown in Figure L8.3.

Argumentation Session

The argumentation session allows all of the groups to share their arguments. One member of each group stays at the lab station to share that group's argument, while the other members of the group go to the other lab stations one at a time to listen to and critique the arguments developed by their classmates. The goal of the argumentation session is not to convince others that your argument is the best one; rather, the goal is to identify errors or instances of faulty reasoning in the initial arguments so these mistakes can be fixed. You will therefore need to evaluate the content of the claim, the quality of the evidence used

Solutes and the Freezing Point of Water
How Does the Addition of Different Types of Solutes Affect the Freezing Point of Water?

to support the claim, and the strength of the justification of the evidence included in each argument that you see. To critique an argument, you might need more information than what is included on the whiteboard. You might therefore need to ask the presenter one or more follow-up questions, such as:

- How did your group collect the data? Why did you use that method?
- What did your group do to make sure the data you collected are reliable? What did you do to decrease measurement error?
- What did your group do to analyze the data, and why did you decide to do it that way? Did you check your calculations?
- Is that the only way to interpret the results of your group's analysis? How do you know that your interpretation of the analysis is appropriate?
- Why did your group decide to present your evidence in that manner?
- What other claims did your group discuss before deciding on that one? Why did you abandon those alternative ideas?
- How confident are you that your group's claim is valid? What could you do to increase your confidence?

Once the argumentation session is complete, you will have a chance to meet with your group and revise your original argument. Your group might need to gather more data or design a way to test one or more alternative claims as part of this process. Remember, your goal at this stage of the investigation is to develop the most valid or acceptable answer to the research question!

Report

Once you have completed your research, you will need to prepare an *investigation report* that consists of three sections that provide answers to the following questions:

1. What were you trying to do and why?
2. What did you do during your investigation and why did you conduct your investigation in this way?
3. What is your argument?

Your report should answer these questions in two pages or less. The report must be typed and any diagrams, figures, or tables should be embedded into the document. Be sure to write in a persuasive style; you are trying to convince others that your claim is acceptable or valid!

LAB 8

Checkout Questions

Lab 8. Solutes and the Freezing Point of Water: How Does the Addition of Different Types of Solutes Affect the Freezing Point of Water?

1. Describe the differences between electrolytes and non-electrolytes.

2. Cities and towns that frequently experience snow in the winter will usually keep warehouses full of salt. When the public roads become covered in snow and ice, they send out trucks to pour the salt over those roads.

    Using what you know about electrolytes and physical properties, explain why it is beneficial to put salt on icy roads.

3. Measuring the freezing point of a solution is an example of an experiment.

    a. I agree with this statement.
    b. I disagree with this statement.

    Explain your answer, using an example from your investigation about solutes and the freezing point of water.

Solutes and the Freezing Point of Water
How Does the Addition of Different Types of Solutes Affect the Freezing Point of Water?

4. Scientists need to be creative or have a good imagination to excel in science.

    a. I agree with this statement.
    b. I disagree with this statement.

    Explain your answer, using an example from your investigation about solutes and the freezing point of water.

5. Scientists often need to track how matter moves into and within a system. Explain why this is important, using an example from your investigation about solutes and the freezing point of water.

6. Scientists often look for and attempt to explain patterns in nature. Explain why patterns are important, using an example from your investigation about solutes and the freezing point of water.

Application Labs

LAB 9

Lab Handout

Lab 9. Melting and Freezing Points: Why Do Substances Have Specific Melting and Freezing Points?

Introduction

All molecules are constantly in motion, so they have *kinetic energy*. The amount of kinetic energy a molecule has depends on its velocity and its mass. *Temperature* is a measure of the average kinetic energy of all the molecules found within a substance. The temperature of a substance, as a result, will go up when the average kinetic energy of the molecules found within that substance increases, and the temperature of a substance will go down when the average kinetic energy of these molecules decreases. Temperature is therefore a useful way to measure how energy moves into, through, or out of a substance.

A substance will change state when enough energy transfers into or out of it. For example, when a solid gains enough energy from its surroundings, it will change into a liquid. This process is called *melting*. Melting is an endothermic process because the substance absorbs energy. The opposite occurs when a substance changes from a liquid to a solid (i.e., *freezing*). When a liquid substance releases enough energy into its surroundings, it will turn into a solid. Freezing is therefore an *exothermic* process because the substance releases energy. All substances have a specific melting point and a specific *freezing point*. The melting point is the temperature at which a substance transitions from a solid to a liquid. The freezing point is the temperature at which a substance transitions from a liquid to a solid.

There are three important properties of a substance that will affect its melting or freezing point. The first property is the molar mass of the molecules that make up that substance. The second property is the nature of the intermolecular forces that hold the molecules of a substance together. The third property is the shape of the molecules that make up that substance.

There are various types of intermolecular forces with different amounts of strength. One type of intermolecular force is caused by *dipole-dipole* interactions. This type of intermolecular force occurs when slightly negative atoms within a molecule are attracted to the slightly positive atoms in nearby molecules. *Hydrogen bonding* is a type of dipole-dipole interaction, in which oxygen, nitrogen, or fluorine atoms in a molecule are strongly attracted to hydrogen atoms in neighboring molecules. Another type of intermolecular force is caused when the arrangement of electrons within an atom has a momentary distribution that is not symmetrical. An asymmetrical distribution of electrons results in the molecule having a temporary dipole. This type of intermolecular interaction, called *London dispersion forces*, tends to occur in nonpolar molecules. London dispersion forces are much weaker than other intermolecular forces such as hydrogen bonding and other dipole-dipole interactions. Even

Melting and Freezing Points
Why Do Substances Have Specific Melting and Freezing Points?

though the relative strength of London dispersion forces is low, the influence of this type of intermolecular force can be magnified when large molecules interact with each other because the opportunity for these types of forces to occur increases.

In this investigation, you will explore how the three properties listed earlier in this section—molar mass, intermolecular forces, and shape of molecules—are related to the specific melting or freezing point of a substance.

Your Task

Determine the melting or freezing point of water, lauric acid, oleic acid, and stearic acid. Then develop a conceptual model that can be used to explain the observed differences in the melting or freezing points of these four substances. Your model should also be able to explain the shape of a heating or cooling curve for each of these substances. Once you have developed your conceptual model, you will need to test it to determine if it allows you to predict the melting point of palmitic acid.

The guiding question of this investigation is, **Why do substances have specific melting and freezing points?**

Materials

You may use any of the following materials during your investigation:

Consumables	Equipment
• Ice	• Temperature sensor kit
• Rock salt	• Hot plate
• Lauric acid, $C_{12}H_{24}O_2$	• Test tubes
• Oleic acid, $C_{18}H_{34}O_2$	• Beaker (500 ml)
• Palmitic acid, $C_{16}H_{32}O_2$	• Styrofoam cup
• Stearic acid, $C_{18}H_{36}O_2$	
• Distilled water	

Safety Precautions

Follow all normal lab safety rules. Your teacher will explain relevant and important information about working with the chemicals associated with this investigation. In addition, take the following safety precautions:

- Wear indirectly vented chemical-splash goggles and chemical-resistant gloves and apron while in the laboratory.
- Handle all glassware with care.
- Use caution when working with hot plates because they can burn skin. Hot plates also need to be kept away from water and other liquids.
- Wash your hands with soap and water before leaving the laboratory.

LAB 9

Investigation Proposal Required?   ☐ Yes   ☐ No

Getting Started

The first step in developing your model is to determine the melting point or the freezing point of each substance using the temperature sensor kit. To conduct these tests, you must determine what type of data you need to collect, how you will collect the data, and how you will analyze the data.

To determine *what type of data you need to collect*, think about the following questions:

- What type of measurements or observations will you need to record during each test?
- When will you need to make these measurements or observations?

To determine *how you will collect the data*, think about the following questions:

- What factors will you control during your different tests?
- What types of comparisons will you need to make?
- What will you do to reduce measurement error?
- How will you keep track of the data you collect and how will you organize it?

To determine *how you will analyze the data*, think about the following questions:

- What type of calculations will you need to make?
- What type of graph could you create to help make sense of your data?
- How will you determine the melting or freezing point of the different substances?
- How many different factors do you need to include in your explanatory model?

Once you have determined the melting or freezing points of each substance, your group will need to develop your conceptual model. The model must be able to explain the differences in melting point that you observe in each substance and the shape of the heating or cooling curves you generated. Your model also must include information about what is taking place between molecules on the submicroscopic level. To develop your model, you will need to consider how the molecular characteristics of each substance, such as its molecular structure or shape and its molar mass, may or may not affect the melting or freezing point of each substance. The table below lists the molar mass and structural formula of each substance for comparison purposes. You can also go to the Chemical Education Digital Library (*www.chemeddl.org*) to learn more about these substances and to view 3-D interactive virtual models of each one.

The last step in this investigation is to test your model. To accomplish this goal, you can use palmitic acid to determine if your model enables you to make accurate predictions

Melting and Freezing Points
Why Do Substances Have Specific Melting and Freezing Points?

about the melting or freezing point of a different substance. The molar mass and structural formula of palmitic acid is provided in Table L9.1. If you are able to use your model to make accurate predictions, then you will be able to generate the evidence you need to convince others that your model is valid.

TABLE L9.1
Molar mass and structural formulas for the substances used in this investigation

Substance	Molar mass	Structural formula
Water	18.02 g/mol	H–Ö–H
Lauric acid	200.32 g/mol	CH₃(CH₂)₁₀COOH (structural formula shown)
Palmitic acid	256.42 g/mol	CH₃(CH₂)₁₄COOH (structural formula shown)
Oleic acid	282.46 g/mol	HOOC(CH₂)₇CH=CH(CH₂)₇CH₃ (structural formula shown)
Stearic acid	284.48 g/mol	CH₃(CH₂)₁₆COOH (structural formula shown)

Connections to Crosscutting Concepts, the Nature of Science, and the Nature of Scientific Inquiry

As you work through your investigation, be sure to think about

- how models are used to help understand natural phenomena,
- how energy and matter flow within systems,
- the different types of investigations conducted in science, and
- the role imagination and creativity play in science.

Initial Argument

Once your group has finished collecting and analyzing your data, you will need to develop an initial argument. Your argument must include a *claim*, which is your answer to the guiding question. Your argument must also include *evidence* in support of your claim. The evidence is your analysis of the data and your interpretation of what the analysis means. Finally, you must include a *justification* of the evidence in your argument. You

LAB 9

FIGURE L9.1
Argument presentation on a whiteboard

The Guiding Question:	
Our Claim:	
Our Evidence:	Our Justification of the Evidence:

will therefore need to use a scientific concept or principle to explain why the evidence that you decided to use is relevant and important. You will create your initial argument on a whiteboard. Your whiteboard must include all the information shown in Figure L9.1.

Argumentation Session

The argumentation session allows all of the groups to share their arguments. One member of each group stays at the lab station to share that group's argument, while the other members of the group go to the other lab stations one at a time to listen to and critique the arguments developed by their classmates. The goal of the argumentation session is not to convince others that your argument is the best one; rather, the goal is to identify errors or instances of faulty reasoning in the initial arguments so these mistakes can be fixed. You will therefore need to evaluate the content of the claim, the quality of the evidence used to support the claim, and the strength of the justification of the evidence included in each argument that you see. To critique an argument, you might need more information than what is included on the whiteboard. You might therefore need to ask the presenter one or more follow-up questions, such as:

- What did your group do to analyze the data, and why did you decide to do it that way?
- Is that the only way to interpret the results of your group's analysis? How do you know that your interpretation of the analysis is appropriate?
- Why did your group decide to present your evidence in that manner?
- What other claims did your group discuss before deciding on that one? Why did you abandon those alternative ideas?
- How confident are you that your group's claim is valid? What could you do to increase your confidence?

Once the argumentation session is complete, you will have a chance to meet with your group and revise your original argument. Your group might need to gather more data or design a way to test one or more alternative claims as part of this process. Remember, your goal at this stage of the investigation is to develop the most valid or acceptable answer to the research question!

Report

Once you have completed your research, you will need to prepare an *investigation report* that consists of three sections that provide answers to the following questions:

Melting and Freezing Points
Why Do Substances Have Specific Melting and Freezing Points?

1. What question were you trying to answer and why?
2. What did you do during your investigation and why did you conduct your investigation in this way?
3. What is your argument?

Your report should answer these questions in two pages or less. The report must be typed and any diagrams, figures, or tables should be embedded into the document. Be sure to write in a persuasive style; you are trying to convince others that your claim is acceptable or valid!

Checkout Questions

Lab 9. Melting and Freezing Points: Why Do Substances Have Specific Melting and Freezing Points?

1. Describe the differences between dipole-dipole interactions and London dispersion forces.

2. Jake is cooking dinner for his family. In order to cook the different parts of his meal he needs to heat one pot of water to boiling (100°C) and another pot of vegetable oil until it reaches about 150°C.

    Use what you know about dipole-dipole interactions, London dispersion forces, molecular mass, and energy transfer to explain how the vegetable oil can reach a higher temperature than the water before it begins to boil.

Melting and Freezing Points
Why Do Substances Have Specific Melting and Freezing Points?

3. In science, experiments are the best way to investigate any question.

   a. I agree with this statement.
   b. I disagree with this statement.

   Explain your answer, using an example from your investigation about melting and freezing points.

4. A painter can be creative during her work, but there is no room for creativity when conducting scientific investigations.

   a. I agree with this statement.
   b. I disagree with this statement.

   Explain your answer, using an example from your investigation about melting points and freezing points.

LAB 9

5. Understanding how matter and energy flow within and between systems is important in science. Explain why this is important, using an example from your investigation about melting and freezing points.

6. Scientists rely on models when they are trying to explain or understand complex phenomena. Explain what a model is and why models are important, using an example from your investigation about melting and freezing points.

Lab Handout

Lab 10. Identification of an Unknown Based on Physical Properties: What Type of Solution Is the Unknown Liquid?

Introduction

Physical changes to chemical substances occur when the physical appearance and structure of a substance change but the chemical structure does not. *Chemical changes* happen when the chemical structure of matter is altered into a different structure that is a completely different substance. Chemists can understand which kind of changes have occurred by examining their relevant physical and chemical properties. *Physical properties* of matter are characteristics of substances that relate to the structure of matter and can be measured without changing them. These properties can be observed or measured in a variety of ways. Physical properties can include boiling point, color, conductivity, density, freezing point, melting point, viscosity, and many more. *Chemical properties* refer to characteristics of matter that can only be observed by changing the structure of a substance through a chemical change. Examples of chemical properties include reactivity, flammability, and oxidation states.

Many of the liquids that you interact with daily are aqueous solutions (see Figure L10.1, p. 84). Dissolving a solute in water can result in either a physical change or a chemical change, depending on the chemical properties of the solute. If a solute can separate into different chemical substances when mixed in water, then a chemical change has occurred. But if the solute stays in the same chemical structure when mixed with water, then only a physical change has occurred because the solute has not changed its chemical structure; it has only changed its physical distribution in the water. The differences between these changes affect the *colligative properties* of the solutions. Colligative properties of solutions are characteristics that are dependent on the number of particles present and not on the specific type of particles present in a solution.

LAB 10

Your Task

Determine the identity of an unknown solution by comparing its physical and chemical properties with the same properties of known solutions.

The guiding question of this investigation is, **What type of solution is the unknown liquid?**

FIGURE L10.1

A sample of methyl blue (left) and an aqueous solution of methyl blue (right)

Materials

You may use any of the following materials during your investigation:

Consumables	Equipment
• 1 M solution of acetic acid, CH_3COOH • 2 M solution of CH_3COOH • 1 M solution of sodium chloride, NaCl • 2 M solution of NaCl • 1 M solution of sucrose, $C_{12}H_{22}O_{11}$ • 2 M solution of $C_{12}H_{22}O_{11}$ • Unknown solution A • Unknown solution B	• Erlenmeyer flasks • Temperature probe • Hot plate • Electronic or triple beam balance • Graduated cylinder (50 ml) • Conductivity meter • pH paper

Safety Precautions

Follow all normal lab safety rules. Your teacher will explain relevant and important information about working with the chemicals associated with this investigation. In addition, take the following safety precautions:

- Wear indirectly vented chemical-splash goggles and chemical-resistant gloves and apron while in the laboratory.

Identification of an Unknown Based on Physical Properties
What Type of Solution Is the Unknown Liquid?

- Use caution when working with hot plates because they can burn skin. Hot plates also need to be kept away from water and other liquids.
- Handle all glassware with care.
- Wash your hands with soap and water before leaving the laboratory.

Investigation Proposal Required?   ☐ Yes   ☐ No

Getting Started

The first step in this investigation is to identify all the various physical and chemical properties that are possible to measure using the available materials. Once you have determined which physical and chemical properties you can measure, you can then design your investigation. To do this, you will need to think about what type of data you need to collect, how you will collect the data, and how you will analyze the data. To determine *what type of data you need to collect*, think about what type of measurements you will need to record during your investigation.

To determine *how you will collect the data*, think about the following questions:

- What will serve as a control (or comparison) condition?
- What types of treatment conditions will you need to set up and how will you do it?
- How will you make sure that your data are of high quality (i.e., how will you reduce error)?

To determine *how you will analyze the data*, think about the following questions:

- How will you determine if there is a difference between the treatment and control conditions?
- What type of table could you create to help make sense of your data?

Connections to Crosscutting Concepts, the Nature of Science, and the Nature of Scientific Inquiry

As you work through your investigation, be sure to think about

- how scale, proportion, and quantity are important in understanding the natural world;
- how energy and matter are related to each other;
- the difference between observations and inferences; and
- the difference between data and evidence.

LAB 10

FIGURE L10.2
Argument presentation on a whiteboard

The Guiding Question:	
Our Claim:	
Our Evidence:	Our Justification of the Evidence:

Initial Argument

Once your group has finished collecting and analyzing your data, you will need to develop an initial argument. Your argument must include a *claim*, which is your answer to the guiding question. Your argument must also include *evidence* in support of your claim. The evidence is your analysis of the data and your interpretation of what the analysis means. Finally, you must include a *justification* of the evidence in your argument. You will therefore need to use a scientific concept or principle to explain why the evidence that you decided to use is relevant and important. You will create your initial argument on a whiteboard. Your whiteboard must include all the information shown in Figure L10.2.

Argumentation Session

The argumentation session allows all of the groups to share their arguments. One member of each group stays at the lab station to share that group's argument, while the other members of the group go to the other lab stations one at a time to listen to and critique the arguments developed by their classmates. The goal of the argumentation session is not to convince others that your argument is the best one; rather, the goal is to identify errors or instances of faulty reasoning in the initial arguments so these mistakes can be fixed. You will therefore need to evaluate the content of the claim, the quality of the evidence used to support the claim, and the strength of the justification of the evidence included in each argument that you see. To critique an argument, you might need more information than what is included on the whiteboard. You might therefore need to ask the presenter one or more follow-up questions, such as:

- What did your group do to make sure the data you collected are reliable? What did you do to decrease measurement error?
- What did your group do to analyze the data, and why did you decide to do it that way? Did you check your calculations?
- Is that the only way to interpret the results of your group's analysis? How do you know that your interpretation of the analysis is appropriate?
- Why did your group decide to present your evidence in that manner?
- What other claims did your group discuss before deciding on that one? Why did you abandon those alternative ideas?
- How confident are you that your group's claim is valid? What could you do to increase your confidence?

Identification of an Unknown Based on Physical Properties
What Type of Solution Is the Unknown Liquid?

Once the argumentation session is complete, you will have a chance to meet with your group and revise your original argument. Your group might need to gather more data or design a way to test one or more alternative claims as part of this process. Remember, your goal at this stage of the investigation is to develop the most valid or acceptable answer to the research question!

Report

Once you have completed your research, you will need to prepare an *investigation report* that consists of three sections that provide answers to the following questions:

1. What question were you trying to answer and why?
2. What did you do during your investigation and why did you conduct your investigation in this way?
3. What is your argument?

Your report should answer these questions in two pages or less. The report must be typed and any diagrams, figures, or tables should be embedded into the document. Be sure to write in a persuasive style; you are trying to convince others that your claim is acceptable or valid!

Checkout Questions

Lab 10. Identification of an Unknown Based on Physical Properties: What Type of Solution Is the Unknown Liquid?

1. What are colligative properties?

2. Tree sap is the liquid that flows within trees to carry nutrients to different parts of the tree. Tree sap is primarily water mixed with other sugars from the tree; this mixture forms a thick syrup-like substance. For trees that grow in very cold climates, where winter temperatures drop below 0°C, it is important that their sap contain a lot of dissolved sugars; otherwise, the trees could die because their sap would freeze.

    Use what you know about colligative properties to explain why tree sap with a lot of dissolved sugars is beneficial to trees living in cold climates.

Identification of an Unknown Based on Physical Properties
What Type of Solution Is the Unknown Liquid?

3. "The freezing point of the solution is –3°C" is an example of an observation.

    a. I agree with this statement.
    b. I disagree with this statement.

    Explain your answer, using an example from your investigation about identification of an unknown based on physical properties.

4. "The freezing points of solutions A and B are both –1°C" is an example of evidence.

    a. I agree with this statement.
    b. I disagree with this statement.

    Explain your answer, using an example from your investigation about identification of an unknown based on physical properties.

LAB 10

5. Scientists often need to look for proportional relationships between different quantities during an investigation. Explain what a proportional relationship is and why these relationships are important, using an example from your investigation about identification of an unknown based on physical properties.

6. It is often important to track how matter flows into, out of, and within system during an investigation. Explain why it is important to keep track of matter when studying a system, using an example from your investigation about identification of an unknown based on physical properties.

Lab Handout

Lab 11. Atomic Structure and Electromagnetic Radiation: What Are the Identities of the Unknown Powders?

Introduction

According to our current theory about the structure of atoms, electrons are found around the nucleus in regions called orbitals (see Figure 11.1). Orbitals represent the potential position of an electron at any given point in time. Orbitals are located at different distances from the nucleus and have different energy levels associated with them. Each orbital, however, can only hold two electrons. The electrons of an atom fill low-energy orbitals, which are the ones closer to the nucleus, before they fill higher-energy ones.

Electrons are in a ground state when under stable conditions. When the electrons in an atom are bombarded with energy from an outside source, however, they absorb that energy and jump temporarily to a higher energy level. The electrons are said to be in an excited state when this happens. When those electrons release that energy, it is emitted in the form of electromagnetic radiation. If that electromagnetic radiation falls between 400 and 700 nanometers (nm) in wavelength, it is given off in the form of visible light.

Many common metal ions, such as Li^+, Na^+, K^+, Ca^{2+}, Ba^{2+}, Sr^{2+}, and Cu^{2+}, produce a distinct color of visible light when they are heated. These ions emit a unique color of light because they consist of atoms that have a unique electron configuration. Chemists can therefore identify these elements with a flame test. To conduct a flame test, a clean wire loop or a wooden splint that has been soaked in distilled water is dipped into a powder or solution and then placed into the hottest portion of a flame (see Figure L11.2).

FIGURE L11.1
Each of the three p orbitals (top row) and all three together on the same atom (bottom)

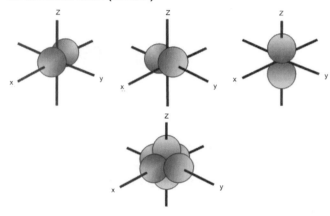

FIGURE L11.2
Flame test

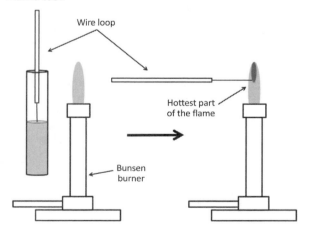

LAB 11

The unique color that we observe during a flame test is actually a mixture of several different wavelengths of visible light. Chemists can use a spectroscope to identify these various wavelengths. This technique is known as spectroscopy. A spectroscope splits light to form an emission line spectrum. The emission line spectrum for hydrogen is provided in Figure L11.3. The emission line spectrum for hydrogen consists of four different wavelengths of light (410 nm, 434 nm, 486 nm, and 656 nm). In this investigation, you will have an opportunity to conduct a flame test and use a spectroscope to identify four unknown powders.

FIGURE L11.3

The hydrogen emission spectrum with wavelength labels

Your Task

Use a flame test and a spectroscope to determine the emission line spectrum of six different powders. Then determine the identity of four unknown powders using a flame test, a spectroscope, and the emission line spectra from the six known powders.

The guiding question of this investigation is, **What are the identities of the unknown powders?**

Materials

You may use any of the following materials during your investigation:

Consumables	Equipment
• Calcium chloride, $CaCl_2$	• Beakers
• Copper(II) chloride, $CuCl_2$	• Bunsen burner
• Lithium chloride, LiCl	• Wooden splints
• Potassium chloride, KCl	• Spectroscope
• Sodium chloride, NaCl	
• Strontium chloride, $SrCl_2$	
• 4 unknown powders	

Safety Precautions

Follow all normal lab safety rules. Your teacher will explain relevant and important information about working with the chemicals associated with this investigation. In addition, take the following safety precautions:

Atomic Structure and Electromagnetic Radiation
What Are the Identities of the Unknown Powders?

- Wear indirectly vented chemical-splash goggles and chemical-resistant gloves and apron while in the laboratory.
- Use caution when working with Bunsen burners. They can burn skin, and combustibles and flammables must be kept away from the open flame. If you have long hair, tie it back behind your head.
- Handle all glassware with care.
- Wash your hands with soap and water before leaving the laboratory.

Investigation Proposal Required?   ☐ Yes   ☐ No

Getting Started

To answer the guiding question, you will need to design and conduct an investigation. To accomplish this task, you must determine what type of data you need to collect, how you will collect the data, and how you will analyze the data.

To determine *what type of data you need to collect*, think about the following questions:

- How will you be able to identify a substance based on a flame test?
- What type of measurements or observations will you need to record during your investigation?

To determine *how you will collect the data*, think about the following questions:

- How often will you collect data and when will you do it?
- How will you make sure that your data are of high quality (i.e., how will you reduce error)?
- How will you keep track of the data you collect and how will you organize it?

To determine *how you will analyze the data*, think about the following questions:

- What type of data table could you create to help make sense of your data?
- What types of calculations will you need to make?

Connections to Crosscutting Concepts, the Nature of Science, and the Nature of Scientific Inquiry

As you work through your investigation, be sure to think about

- the importance of identifying patterns,
- how system models contribute to understanding science,
- the difference between laws and theories in science, and
- the importance of imagination and creativity in your investigation.

LAB 11

Initial Argument

Once your group has finished collecting and analyzing your data, you will need to develop an initial argument. Your argument must include a *claim*, which is your answer to the guiding question. Your argument must also include *evidence* in support of your claim. The evidence is your analysis of the data and your interpretation of what the analysis means. Finally, you must include a *justification* of the evidence in your argument. You will therefore need to use a scientific concept or principle to explain why the evidence that you decided to use is relevant and important. You will create your initial argument on a whiteboard. Your whiteboard must include all the information shown in Figure L11.4.

FIGURE L11.4
Argument presentation on a whiteboard

The Guiding Question:	
Our Claim:	
Our Evidence:	Our Justification of the Evidence:

Argumentation Session

The argumentation session allows all of the groups to share their arguments. One member of each group stays at the lab station to share that group's argument, while the other members of the group go to the other lab stations one at a time to listen to and critique the arguments developed by their classmates. The goal of the argumentation session is not to convince others that your argument is the best one; rather, the goal is to identify errors or instances of faulty reasoning in the initial arguments so these mistakes can be fixed. You will therefore need to evaluate the content of the claim, the quality of the evidence used to support the claim, and the strength of the justification of the evidence included in each argument that you see. To critique an argument, you might need more information than what is included on the whiteboard. You might therefore need to ask the presenter one or more follow-up questions, such as:

- How did your group collect the data? Why did you use that method?
- What did your group do to make sure the data you collected are reliable? What did you do to decrease measurement error?
- What did your group do to analyze the data, and why did you decide to do it that way?
- Is that the only way to interpret the results of your group's analysis? How do you know that your interpretation of the analysis is appropriate?
- Why did your group decide to present your evidence in that manner?
- What other claims did your group discuss before deciding on that one? Why did you abandon those alternative ideas?
- How confident are you that your group's claim is valid? What could you do to increase your confidence?

Atomic Structure and Electromagnetic Radiation
What Are the Identities of the Unknown Powders?

Once the argumentation session is complete, you will have a chance to meet with your group and revise your original argument. Your group might need to gather more data or design a way to test one or more alternative claims as part of this process. Remember, your goal at this stage of the investigation is to develop the most valid or acceptable answer to the research question!

Report

Once you have completed your research, you will need to prepare an *investigation report* that consists of three sections that provide answers to the following questions:

1. What question were you trying to answer and why?
2. What did you do during your investigation and why did you conduct your investigation in this way?
3. What is your argument?

Your report should answer these questions in two pages or less. The report must be typed and any diagrams, figures, or tables should be embedded into the document. Be sure to write in a persuasive style; you are trying to convince others that your claim is acceptable or valid!

LAB 11

Checkout Questions

Lab 11. Atomic Structure and Electromagnetic Radiation: What Are the Identities of the Unknown Powders?

1. Describe how photons can be emitted from an atom.

2. Neon lights used in many signs and commercial displays give off a unique red light. These lights work by passing an electric current through a glass tube filled with neon gas (a noble gas). Neon gas produces a red color, however "neon" lights can be found in a variety of other colors too.

    Use what you know about photons and atomic structure to explain how it is possible to produce other colors of "neon" light.

3. Theories and laws are different kinds of scientific knowledge.

    a. I agree with this statement.//
    b. I disagree with this statement.

    Explain your answer, using an example from your investigation about atomic structure and electromagnetic radiation.

Atomic Structure and Electromagnetic Radiation
What Are the Identities of the Unknown Powders?

4. Scientists need to be creative or have a good imagination to excel in science.

   a. I agree with this statement.

   b. I disagree with this statement.

   Explain your answer, using an example from your investigation about atomic structure and electromagnetic radiation.

5. Scientists often use models to help them understand natural phenomena. Explain what a model is and why models are important, using an example from your investigation about atomic structure and electromagnetic radiation.

6. Scientists often look for and attempt to explain patterns in nature. Explain why patterns are important, using an example from your investigation about atomic structure and electromagnetic radiation.

LAB 12

Lab Handout

Lab 12. Magnetism and Atomic Structure: What Relationships Exist Between the Electrons in a Substance and the Strength of Magnetic Attraction?

Introduction

The structure of atoms is both simple and complex at the same time. The bulk of the mass of an atom is located in the nucleus, where protons and neutrons are relatively evenly distributed. Outside an atom's nucleus are electrons, which are negatively charged particles that are attracted to the positively charged nucleus. Electrons inhabit regions of space known as orbitals. These regions represent the probable position of an electron at any given point in time. Each orbital has a unique shape and can hold only two electrons. Each orbital also has a defined energy level.

At the first energy level is the 1s orbital. The "1" indicates that the orbital is in the energy level closest to the nucleus, and the "s" describes the shape of the orbital; s orbitals are spheres. At the second energy level, there is a second s orbital called the 2s orbital. This orbital is similar to the 1s orbital except that the region where there is the greatest chance of finding the electron is farther out from the nucleus. There are similar orbitals at higher energy levels (e.g., 3s, 4s, and 5s). At the second energy level there are also p orbitals; a p orbital looks like two balloons tied together. There are three different p orbitals that point at right angles to each other. These orbitals are called p_x, p_y, and p_z. The p orbitals at the second energy level are called $2p_x$, $2p_y$, and $2p_z$ (the 2 indicates that they are at the second energy level). There are also p orbitals at higher energy levels (e.g., $3p_x$, $3p_y$, $3p_z$, $4p_x$, $4p_y$, and $4p_z$). In addition to s and p orbitals, there are two other sets of orbitals at higher energy levels. At the third energy level, for example, there is a set of five d orbitals (with complicated shapes and names) as well as the 3s and 3p orbitals. At the third energy level there are a total of nine orbitals. At the fourth energy level, there are seven f orbitals as well the 4s, 4p, and 4d orbitals. Figure L12.1 illustrates the 1s, 2s, and 2p orbitals.

The electron configuration of any atom can be determined using the *Aufbau principle*. According to this principle, electrons fill low-energy orbitals before they fill higher-energy ones. Electrons, however, are mutually repulsive, so individual electrons will occupy different orbitals at the same energy level before sharing the same orbital at a specific energy level. This tendency to occupy different orbitals at the same energy level, which is called Hund's rule, helps to minimize the repulsions between electrons and makes the atom more stable. As an example, consider the electron configuration of iron as illustrated in Figure L12.2. In this figure orbitals are represented as horizontal lines with the electrons

Magnetism and Atomic Structure
What Relationships Exist Between the Electrons in a Substance and the Strength of Magnetic Attraction?

FIGURE L12.1
3-D representation of s and p electron orbitals

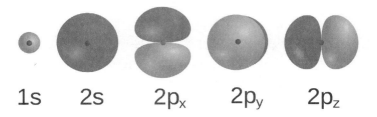

FIGURE L12.2
Electron configuration diagram for iron

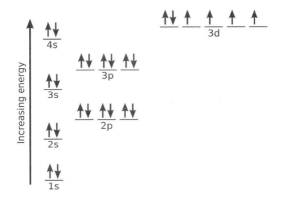

in them depicted as arrows; up and down arrows are used to indicate that the electrons have different spins. The 26 electrons that are present in an iron atom are distributed across 15 orbitals according to the Aufbau principle and Hund's rule. The 1s, 2s, 2p, 3s, and 4s orbitals are completely filled. The five 3d orbitals are each occupied by at least one electron, and only one of these orbitals is completely filled.

The electron configuration of the atoms found within a substance will affect its magnetic properties. Many pure substances, besides iron metal, are attracted to a strong magnetic field. These substances are said to be paramagnetic. Substances that are not affected by strong magnetic fields are said to be diamagnetic. Past experiments with electricity and magnetism have demonstrated a connection between the magnetic properties of a substance and the movement of electrons. In the quantum mechanical model, one of the energy states of electrons is related to the electron's spin ($m_s = -1/2$ or $+1/2$). This electron spin produces a magnetic field with electrons spinning in opposite directions that have reverse magnetic poles. Paired electrons always have the opposite spin from their partner, which means their magnetic fields cancel out each other's effect. Unpaired electrons can spin in either direction and can change that direction when interacting with the magnetic

LAB 12

fields of other unpaired electrons. In this investigation you will explore the relationship between electron configuration and magnetism in ionic compounds.

Your Task

Determine how electron configuration affects the magnetic properties of a substance.

The guiding question of this investigation is, **What relationships exist between the electrons in a substance and the strength of magnetic attraction?**

Materials

You may use any of the following materials during your investigation:

Consumables	Equipment
Vials containing copper(II) chloride (Cu^{2+}), iron(II) chloride (Fe^{2+}), manganese(II) chloride (Mn^{2+}), and zinc chloride (Zn^{2+})	• Empty vial • Neodymium-iron-boron magnet • Electronic balance • Plastic collar for electronic balance plate • Splints or rods

Safety Precautions

Follow all normal lab safety rules. Your teacher will explain relevant and important information about working with the chemicals associated with this investigation. *Caution: The neodymium-iron-boron magnet that you will use in this investigation is very powerful but extremely brittle.* **DO NOT** *test the magnet on the exposed metal at your laboratory station or bring anything metallic near your magnet.* In addition, take the following safety precautions:

- Wear indirectly vented chemical-splash goggles and chemical-resistant gloves and apron while in the laboratory.
- Wash your hands with soap and water before leaving the laboratory.

Investigation Proposal Required?   ☐ Yes   ☐ No

Getting Started

The first step in your investigation will be to create a device that you can use to measure the magnetic properties of different substances. Figure L12.3 shows how you can use an electronic balance to measure magnetic attraction. Once you have created your magnetic attraction measuring device, you will need to develop a procedure to determine how electron configuration affects the magnetic properties of the different chlorides that are available. Before you can design your procedure, however, your group will need to determine what type of data you need to collect, how you will collect the data, and how you will analyze the data.

Magnetism and Atomic Structure
What Relationships Exist Between the Electrons in a Substance and the Strength of Magnetic Attraction?

To determine *what type of data you need to collect*, think about the following questions:

- What type of measurements will you need to record during your investigation?
- When will you need to make these measurements?

To determine *how you will collect the data*, think about the following questions:

- What comparisons will you need to make?
- How will you hold other variables constant?
- How will you use reduce measurement error?
- How will you keep track of the data you collect and how you will organize it?

To determine *how you will analyze the data*, think about the following questions:

- What are the similarities and differences in the electron configurations for each of the metal ions you used?
- What type of graph or table could you create to help make sense of your data?

FIGURE L12.3
Magnetic attraction measuring device

Place the vial on the support directly over the magnet

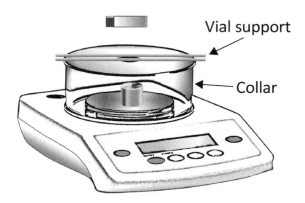

Vial support

Collar

LAB 12

Connections to Crosscutting Concepts, the Nature of Science, and the Nature of Scientific Inquiry

As you work through your investigation, be sure to think about

- the relationship between structure and function,
- how system models contribute to understanding science,
- the difference between laws and theories in science, and
- how scientific knowledge can change over time in light of new evidence.

Initial Argument

Once your group has finished collecting and analyzing your data, you will need to develop an initial argument. Your argument must include a *claim*, which is your answer to the guiding question. Your argument must also include *evidence* in support of your claim. The evidence is your analysis of the data and your interpretation of what the analysis means. Finally, you must include a *justification* of the evidence in your argument. You will therefore need to use a scientific concept or principle to explain why the evidence that you decided to use is relevant and important. You will create your initial argument on a whiteboard. Your whiteboard must include all the information shown in Figure L12.4.

FIGURE L12.4
Argument presentation on a whiteboard

The Guiding Question:	
Our Claim:	
Our Evidence:	Our Justification of the Evidence:

Argumentation Session

The argumentation session allows all of the groups to share their arguments. One member of each group stays at the lab station to share that group's argument, while the other members of the group go to the other lab stations one at a time to listen to and critique the arguments developed by their classmates. The goal of the argumentation session is not to convince others that your argument is the best one; rather, the goal is to identify errors or instances of faulty reasoning in the initial arguments so these mistakes can be fixed. You will therefore need to evaluate the content of the claim, the quality of the evidence used to support the claim, and the strength of the justification of the evidence included in each argument that you see. To critique an argument, you might need more information than what is included on the whiteboard. You might therefore need to ask the presenter one or more follow-up questions, such as:

- How did your group collect the data? Why did you use that method?
- What did your group do to make sure the data you collected are reliable? What did you do to decrease measurement error?

Magnetism and Atomic Structure
What Relationships Exist Between the Electrons in a Substance and the Strength of Magnetic Attraction?

- What did your group do to analyze the data, and why did you decide to do it that way? Did you check your calculations?
- Is that the only way to interpret the results of your group's analysis? How do you know that your interpretation of the analysis is appropriate?
- Why did your group decide to present your evidence in that manner?
- What other claims did your group discuss before deciding on that one? Why did you abandon those alternative ideas?
- How confident are you that your group's claim is valid? What could you do to increase your confidence?

Once the argumentation session is complete, you will have a chance to meet with your group and revise your original argument. Your group might need to gather more data or design a way to test one or more alternative claims as part of this process. Remember, your goal at this stage of the investigation is to develop the most valid or acceptable answer to the research question!

Report

Once you have completed your research, you will need to prepare an *investigation report* that consists of three sections that provide answers to the following questions:

1. What question were you trying to answer and why?
2. What did you do during your investigation and why did you conduct your investigation in this way?
3. What is your argument?

Your report should answer these questions in two pages or less. The report must be typed and any diagrams, figures, or tables should be embedded into the document. Be sure to write in a persuasive style; you are trying to convince others that your claim is acceptable or valid!

Checkout Questions

Lab 12. Magnetism and Atomic Structure: What Relationships Exist Between the Electrons in a Substance and the Strength of Magnetic Attraction?

1. Describe what you know about electron orbitals and how they are filled within an atom.

2. There are various elements that are considered metals. Many metals have shared characteristics, such as being malleable; however, one characteristic that differs among the metals is whether or not the metal is attracted to a magnet. For example, iron is very magnetic, but aluminum is not.

    Use what you know about electron orbital structure and electron spin to explain how some substances are considered paramagnetic (attracted to a strong magnetic field) and other substances are diamagnetic (not affected by a strong magnetic field).

3. Scientific knowledge is a set body of information that does not change.

    a. I agree with this statement.
    b. I disagree with this statement.

    Explain your answer, using an example from your investigation about magnetism and atomic structure.

Magnetism and Atomic Structure
What Relationships Exist Between the Electrons in a Substance and the Strength of Magnetic Attraction?

4. Theories and laws are different kinds of scientific knowledge.

    a. I agree with this statement.
    b. I disagree with this statement.

    Explain your answer, using an example from your investigation about magnetism and atomic structure.

5. Scientists often use models to help them understand natural phenomena. Explain what a model is and why models are important, using an example from your investigation about magnetism and atomic structure.

6. In nature, the structure of an object is often related to function or the properties of that object. Explain why this is true, using an example from your investigation about magnetism and atomic structure.

LAB 13

Lab Handout

Lab 13. Density and the Periodic Table: What Are the Densities of Germanium and Flerovium?

Introduction

At the time Dmitri Mendeleev proposed his periodic table for the classification of the elements in 1869, only 63 elements were known. Mendeleev arranged these 63 elements into a table of rows and columns in order of increasing atomic mass and by repeating physical properties. He also suggested that there were some missing elements that still needed to be discovered.

In Mendeleev's periodic table (see Figure L13.1), carbon and silicon were placed in the same period. Carbon appeared in the 2nd group and silicon appeared in the 3rd group. Mendeleev then proposed that there should be another element in this period. He named the missing element eka-silicium and predicted its physical properties. The German chemist Clemens Winkler discovered a new element in 1886 and called it germanium. In one of his reports of the discovery, he stated,

> It was definitely premature when I expressed such an assumption in my first notice concerning germanium; at least there was no basis for its proof. Nor would I have ventured at first to assume argyrodite to be a sulpho salt with a quadrivalent acid radical, because there were no analogies at all for such an assumption. Thus the present case shows very clearly how treacherous it can be to build upon analogies; the quadrivalency of germanium has by now become an incontrovertible fact, and there can be no longer any doubt that the new element is no other than the eka-silicium prognosticated fifteen years ago by Mendeleev. (Winkler 1886)

As this example from this history of chemistry illustrates, Mendeleev's periodic table gave chemists a powerful tool for predicting the properties of the elements. The periodic table, however, has been reorganized and rearranged over time. Carbon, silicon, and germanium are now placed in group 14 with tin and lead. Group 14 also contains another element called flerovium (atomic number 114). Flerovium is an extremely radioactive element that can only be created in the laboratory. In fact, scientists have only been able to create about 80 atoms of flerovium since it was first produced at the Flerov Laboratory of Nuclear Reactions in 1998. In this investigation, you will use the periodic table to determine the densities of two elements found in group 14.

Density and the Periodic Table
What Are the Densities of Germanium and Flerovium?

FIGURE L13.1
Mendeleev's 1869 periodic table

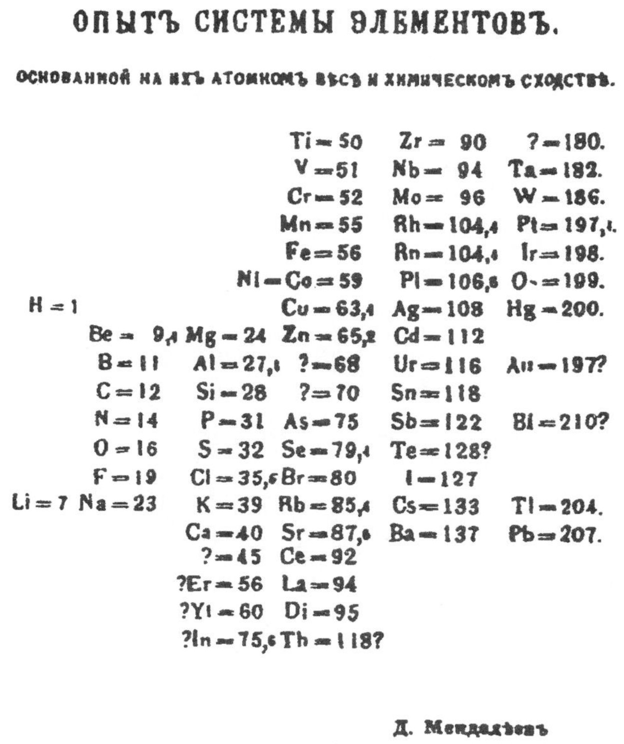

ОПЫТЪ СИСТЕМЫ ЭЛЕМЕНТОВЪ.

ОСНОВАННОЙ НА ИХЪ АТОМНОМЪ ВѢСѢ И ХИМИЧЕСКОМЪ СХОДСТВѢ.

```
                          Ti = 50    Zr = 90    ? = 180.
                          V = 51     Nb = 94    Ta = 182.
                          Cr = 52    Mo = 96    W = 186.
                          Mn = 55    Rh = 104,4 Pt = 197,4.
                          Fe = 56    Rn = 104,4 Ir = 198.
                      Ni = Co = 59   Pl = 106,6 O  = 199.
              H = 1     Cu = 63,4    Ag = 108   Hg = 200.
         Be = 9,4 Mg = 24 Zn = 65,2  Cd = 112
         B = 11   Al = 27,4 ? = 68   Ur = 116   Au = 197?
         C = 12   Si = 28   ? = 70   Sn = 118
         N = 14   P = 31   As = 75   Sb = 122   Bi = 210?
         O = 16   S = 32   Se = 79,4 Te = 128?
         F = 19   Cl = 35,6 Br = 80  I = 127
    Li = 7 Na = 23 K = 39  Rb = 85,4 Cs = 133   Tl = 204.
              Ca = 40 Sr = 87,6 Ba = 137        Pb = 207.
              ? = 45  Ce = 92
          ?Er = 56    La = 94
          ?Yt = 60    Di = 95
          ?In = 75,6 Th = 118?
```

Д. Менделѣевъ

Your Task

Determine the densities of carbon, lead, silicon, and tin by measuring the mass and volume of a sample of each of these elements. Then use this information to predict the densities of the other two elements in group 14 (germanium and flerovium).

The guiding question of this investigation is, **What are the densities of germanium and flerovium?**

Materials

You may use any of the following materials during your investigation:

Consumables	Equipment
• Charcoal	• 4 beakers (each 250 ml)
• Lead shot	• Graduated cylinders (25 ml and 50 ml)
• Silicon shot	• Electronic or triple beam balance
• Tin shot	• Periodic table
• Distilled water	• Weighing dishes

LAB 13

Safety Precautions
Follow all normal lab safety rules. Your teacher will explain relevant and important information about working with the chemicals associated with this investigation. In addition, take the following safety precautions:

- Wear chemical-splash goggles while in the laboratory.
- Handle all glassware with care.
- Wash your hands with soap and water before leaving the laboratory.

Investigation Proposal Required?   ☐ Yes   ☐ No

Getting Started
To answer the guiding question, you will need to think about what type of data you need to collect, how you will collect the data, and how you will analyze the data.

To determine *what type of data you need to collect*, think about the following questions:

- What type of information will you need to collect during your investigation to determine the densities of carbon, lead, silicon, and tin?
- How will you use the densities of carbon, lead, silicon, and tin to predict the densities of the other two elements?

To determine *how you will collect the data*, think about the following questions:

- What equipment will you need?
- How will you reduce measurement error?
- How will you keep track of the data you collect?

To determine *how you will analyze the data*, think about the following questions:

- What types of calculations will you need to make to determine the densities of carbon, lead, silicon, and tin?
- How can you use mathematics to predict the densities of the other two elements based on the densities of carbon, lead, silicon, and tin?
- What type of graph could you create to help make sense of your data?

Connections to Crosscutting Concepts, the Nature of Science, and the Nature of Scientific Inquiry
As you work through your investigation, be sure to think about

- the importance of identifying patterns;

Density and the Periodic Table
What Are the Densities of Germanium and Flerovium?

- issues of scale, proportion, and quantity;
- how scientific knowledge can change over time in light of new evidence; and
- how scientists can use different methods to answer different types of questions.

Initial Argument

Once your group has finished collecting and analyzing your data, you will need to develop an initial argument. Your argument must include a *claim*, which is your answer to the guiding question. Your argument must also include *evidence* in support of your claim. The evidence is your analysis of the data and your interpretation of what the analysis means. Finally, you must include a *justification* of the evidence in your argument. You will therefore need to use a scientific concept or principle to explain why the evidence that you decided to use is relevant and important. You will create your initial argument on a whiteboard. Your whiteboard must include all the information shown in Figure L13.2.

FIGURE L13.2
Argument presentation on a whiteboard

The Guiding Question:	
Our Claim:	
Our Evidence:	Our Justification of the Evidence:

Argumentation Session

The argumentation session allows all of the groups to share their arguments. One member of each group stays at the lab station to share that group's argument, while the other members of the group go to the other lab stations one at a time to listen to and critique the arguments developed by their classmates. The goal of the argumentation session is not to convince others that your argument is the best one; rather, the goal is to identify errors or instances of faulty reasoning in the initial arguments so these mistakes can be fixed. You will therefore need to evaluate the content of the claim, the quality of the evidence used to support the claim, and the strength of the justification of the evidence included in each argument that you see. To critique an argument, you might need more information than what is included on the whiteboard. You might therefore need to ask the presenter one or more follow-up questions, such as:

- How did your group collect the data? Why did you use that method?
- What did your group do to make sure the data you collected are reliable? What did you do to decrease measurement error?
- What did your group do to analyze the data, and why did you decide to do it that way? Did you check your calculations?
- Is that the only way to interpret the results of your group's analysis? How do you know that your interpretation of the analysis is appropriate?
- Why did your group decide to present your evidence in that manner?

- What other claims did your group discuss before deciding on that one? Why did you abandon those alternative ideas?
- How confident are you that your group's claim is valid? What could you do to increase your confidence?

Once the argumentation session is complete, you will have a chance to meet with your group and revise your original argument. Your group might need to gather more data or design a way to test one or more alternative claims as part of this process. Remember, your goal at this stage of the investigation is to develop the most valid or acceptable answer to the research question!

Report

Once you have completed your research, you will need to prepare an *investigation report* that consists of three sections that provide answers to the following questions:

1. What question were you trying to answer and why?
2. What did you do during your investigation and why did you conduct your investigation in this way?
3. What is your argument?

Your report should answer these questions in two pages or less. The report must be typed and any diagrams, figures, or tables should be embedded into the document. Be sure to write in a persuasive style; you are trying to convince others that your claim is acceptable or valid!

Reference

Winkler, C. 1886. About germanium. *Journal für prakische Chemie* 142 (N.F. 34): 177–229. Excerpt available online at *www.chemteam.info/Chem-History/Disc-of-Germanium.html*.

Checkout Questions

Lab 13. Density and the Periodic Table: What Are the Densities of Germanium and Flerovium?

1. What are periodic trends?

2. The following table shows the measured densities for period 2 and group 14 on the periodic table.

Element	Density	Location on the periodic table	
		Period	Group
Lithium	0.53 g/cm³	2	1
Beryllium	1.85 g/cm³	2	2
Boron	2.46 g/cm³	2	13
Carbon	2.26 g/cm³	2	14
Nitrogen	1.25 g/L	2	15
Oxygen	1.43 g/L	2	16
Fluorine	1.70 g/L	2	17
Neon	0.90 g/L	2	18
Silicon	2.33 g/cm³	3	14
Germanium	5.32 g/cm³	4	14
Tin	7.31 g/cm³	5	14
Lead	11.34 g/cm³	6	14

LAB 13

Use what you know about density and periodic trends, along with the data in the table, to explain whether or not density is a periodic trend.

3. Scientific knowledge can change over time in light of new evidence.

   a. I agree with this statement.
   b. I disagree with this statement.

   Explain your answer, using an example from your investigation about density and the periodic table.

4. An investigation must follow the scientific method to be considered scientific.

   a. I agree with this statement.
   b. I disagree with this statement.

   Explain your answer, using an example from your investigation about density and the periodic table.

5. Scientists often look for and attempt to explain patterns in nature. Explain why patterns are important, using an example from your investigation about density and the periodic table.

6. Scientists often need to look for proportional relationships. Explain why looking for a proportional relationship is often useful in science, using an example from your investigation about density and the periodic table.

LAB 14

Lab Handout

Lab 14. Molar Relationships: What Are the Identities of the Unknown Compounds?

Introduction

The concept of the mole is important for understanding chemistry. The mole provides a measure of the number of atoms present in a sample of a compound. One mole of an element or compound contains 6.02×10^{23} atoms or molecules. This quantity is referred to as the Avogadro constant. Knowing the amounts of particles allows chemists to understand how different chemicals behave during chemical reactions and predict the outcomes of reactions. Moles provide a standardized way of comparing elements. Using the Avogadro constant, chemists can use other measures, such as mass or volume, to determine the amount of particles a sample has.

To use mass to determine the number of moles of an element or molecule in a sample, you must also know the molar mass of that element or molecule. The molar mass refers to the total mass of an element present in one mole of that element. The unit for these masses is grams per mole (g/mol). The molar mass of an element is easily identified on most periodic tables, where it is typically listed in the box provided for a particular element. Examples of molar mass include carbon (C), 12.011 g/mol; oxygen (O), 15.994 g/mol; and gold (Au), 196.967 g/mol. To determine the molar mass for a compound made of larger molecules, you must add up the molar masses of all the atoms present in the molecular formula. For example, the molar mass of CO_2 is 43.999 g/mol, which is calculated by 12.011 g/mol (C) + 15.994 g/mol (O) + 15.994 g/mol (O). Remember that you have to include the total number of atoms in the molecular formula when calculating molar mass, so be mindful of the subscripts in those formulas.

By knowing the molar mass of a compound and the mass of a sample of that compound, you can determine the number of moles in the compound. Continuing from the example above, if you have a sample of CO_2 whose mass is 2.523 g, then you can determine the number of moles in that sample by dividing the actual mass by the molar mass (e.g., 2.523 g / 43.999 g/mol = 0.0573 moles of CO_2).

You will now use your understanding of the relationships between moles, molar mass, and mass of a sample to identify some unknown compounds. Remember, moles provide a standardized unit of measure (based on the Avogadro constant) so that chemists can compare a wide variety of substances, including the amount of substances needed and produced by a chemical reaction.

Molar Relationships
What Are the Identities of the Unknown Compounds?

Your Task

You will be given seven sealed bags. Each bag will be filled with a different powder and will be labeled with the number of moles of powder that is inside the bag. Your task will be to identify the powder in each bag. The unidentified powders could be any of the following compounds:

- Calcium acetate, $Ca(C_2H_3O_2)_2$
- Calcium oxide, CaO
- Potassium sulfate, K_2SO_4
- Sodium acetate, $NaC_2H_3O_2$
- Sodium carbonate, Na_2CO_3
- Sodium chloride, $NaCl$
- Zinc(II) oxide, ZnO

The guiding question of this investigation is, **What are the identities of the unknown compounds?**

Materials

You may use any of the following materials during your investigation:

Consumables	Equipment
• Sealed plastic bags of unknown compounds • Empty plastic bags	• Electronic or triple beam balance • Periodic table

Safety Precautions

Follow all normal lab safety rules. Your teacher will explain relevant and important information about working with the chemicals associated with this investigation. In addition, take the following safety precautions:

- Wear indirectly vented chemical-splash goggles while in the laboratory.
- Wash your hands with soap and water before leaving the laboratory.

Investigation Proposal Required?   ☐ Yes   ☐ No

Getting Started

To answer the guiding question, you will need to design and conduct an investigation. To accomplish this task, you must first determine what type of data you need to collect, how you will collect the data, and how you will analyze the data.

To determine *what type of data you need to collect*, think about what type of measurements you will need to make during your investigation.

LAB 14

To determine *how you will collect the data*, think about the following questions:

- How will you make sure that your data are of high quality (i.e., how will you reduce error)?
- How will you keep track of the data you collect and how will you organize it?

To determine *how you will analyze the data*, think about the following questions:

- What type of table or graph could you create to help make sense of your data?
- What types of calculations will you need to make?

Connections to Crosscutting Concepts, the Nature of Science, and the Nature of Scientific Inquiry

As you work through your investigation, be sure to think about

- the importance of identifying patterns,
- which proportional relationships are critical to the understanding of this investigation,
- how scientific knowledge changes over time in light of new evidence, and
- the difference between data and evidence.

Initial Argument

Once your group has finished collecting and analyzing your data, you will need to develop an initial argument. Your argument must include a *claim*, which is your answer to the guiding question. Your argument must also include *evidence* in support of your claim. The evidence is your analysis of the data and your interpretation of what the analysis means. Finally, you must include a *justification* of the evidence in your argument. You will therefore need to use a scientific concept or principle to explain why the evidence that you decided to use is relevant and important. You will create your initial argument on a whiteboard. Your whiteboard must include all the information shown in Figure L14.1.

FIGURE L14.1

Argument presentation on a whiteboard

The Guiding Question:	
Our Claim:	
Our Evidence:	Our Justification of the Evidence:

Argumentation Session

The argumentation session allows all of the groups to share their arguments. One member of each group stays at the lab station to share that group's argument, while the other members of the group go to the other lab stations one at a time to listen to and critique the arguments developed by their classmates. The goal of the argumentation session is not to

Molar Relationships
What Are the Identities of the Unknown Compounds?

convince others that your argument is the best one; rather, the goal is to identify errors or instances of faulty reasoning in the initial arguments so these mistakes can be fixed. You will therefore need to evaluate the content of the claim, the quality of the evidence used to support the claim, and the strength of the justification of the evidence included in each argument that you see. To critique an argument, you might need more information than what is included on the whiteboard. You might therefore need to ask the presenter one or more follow-up questions, such as:

- How did your group collect the data? Why did you use that method?
- What did your group do to make sure the data you collected are reliable? What did you do to decrease measurement error?
- What did your group do to analyze the data? Did you check your calculations?
- Is that the only way to interpret the results of your group's analysis? How do you know that your interpretation of the analysis is appropriate?
- Why did your group decide to present your evidence in that manner?
- What other claims did your group discuss before deciding on that one? Why did you abandon those alternative ideas?
- How confident are you that your group's claim is valid? What could you do to increase your confidence?

Once the argumentation session is complete, you will have a chance to meet with your group and revise your original argument. Your group might need to gather more data or design a way to test one or more alternative claims as part of this process. Remember, your goal at this stage of the investigation is to develop the most valid or acceptable answer to the research question!

Report

Once you have completed your research, you will need to prepare an *investigation report* that consists of three sections that provide answers to the following questions:

1. What question were you trying to answer and why?
2. What did you do during your investigation and why did you conduct your investigation in this way?
3. What is your argument?

Your report should answer these questions in two pages or less. The report must be typed and any diagrams, figures, or tables should be embedded into the document. Be sure to write in a persuasive style; you are trying to convince others that your claim is acceptable or valid!

LAB 14

Checkout Questions

Lab 14. Molar Relationships: What Are the Identities of the Unknown Compounds?

1. A 1-mole sample of sugar ($C_6H_{12}O_6$) is 180 grams, but a 1-mole sample of salt (NaCl) is 58 grams. These two samples are equal when comparing the number of moles, but not equal when comparing mass. Describe why this relationship is possible.

2. The following chemical equation describes the chemical reaction of hydrogen gas and oxygen gas to create water.

$$2H_2 + O_2 \rightarrow 2H_2O$$

    Use what you know about molar relationships to explain how scientists can predict the amount of water produced if they know the amounts of hydrogen and oxygen gases they have to react.

3. Scientific knowledge is a set body of information that can change over time in light of new evidence.

    a. I agree with this statement.

    b. I disagree with this statement.

    Explain your answer, using an example from your investigation about molar relationships.

4. The terms *data* and *evidence* do not have the same meaning in science.

   a. I agree with this statement.

   b. I disagree with this statement.

   Explain your answer, using an example from your investigation about molar relationships.

5. Scientists often look for and attempt to explain patterns in nature. Explain why patterns are important, using an example from your investigation about molar relationships.

6. Scientists often need to look for proportional relationships. Explain why looking for proportional relationships is often useful in science, using an example from your investigation about molar relationships.

LAB 15

Lab Handout

Lab 15. The Ideal Gas Law: How Can a Value of R for the Ideal Gas Law Be Accurately Determined Inside the Laboratory?

Introduction

A *gas* is the state of matter that is characterized by having neither a fixed shape nor a fixed volume. Gases exert pressure, are compressible, have low densities, and diffuse rapidly when mixed with other gases. On a submicroscopic level, the molecules in a gas are separated by large distances and are in constant, random motion. A gas can be described using four measurable properties: pressure (P), defined as the force exerted by a gas per unit area; volume (V), defined as the quantity of space a gas occupies; temperature (T), defined as the average kinetic energy of the molecules that make up a gas; and the number of moles of gas (n). The relationships among these properties are summarized by the gas laws, as shown in Table L15.1.

TABLE L15.1
The gas laws

Gas law	Relationship	Equation
Boyle's law	$V \propto 1/P$ (T and n are held constant). As gas pressure increases, gas volume decreases.	$P_1V_1 = P_2V_2$
Charles' law	$V \propto T$ (P and n are held constant). As gas temperature increases, gas volume increases.	$V_1/T_1 = V_2/T_2$
Gay-Lussac's law	$P \propto T$ (V and n are held constant). As gas pressure increases, gas temperature increases.	$P_1/T_1 = P_2/T_2$
Avogadro's law	$V \propto n$ (P and T are held constant). As the number of moles of gas increase, gas volume increases.	$V_1/n_1 = V_2/n_2$
Combined law	$V \propto T/P$ (n is held constant); obtained by combining Boyle's law, Charles' law, and Gay-Lussac's law.	$(P_1V_1)/T_1 = (P_2V_2)/T_2$

The ideal gas law combines Boyle's law, Charles' law, Gay-Lussac's law, and Avogadro's law to describe the relationship among the pressure, volume, temperature, and number of moles of gas. Émile Clapeyron is often given the credit for developing this law. The ideal gas law provides chemists with a powerful predictive tool that helps them understand how gases will react in different systems. The ideal gas law is expressed mathematically as $PV = nRT$. P is pressure in atmospheres (atm); V is volume in liters (L); n is the number of moles of gas (mol); and T is absolute temperature in Kelvin (K). The remaining component of the ideal gas law is R, which is called the ideal gas constant. The theoretical value of R that is often reported in textbooks and handbooks is 0.0821 L•atm/mol•K or 8.314 L•kPa/mol•K.

As chemists worked to determine an exact value for R in the mid-1800s, they were faced with numerous challenges. First, they had to develop a method that they could use to generate the experimental data they needed to calculate a value for R from the ideal gas law. Second, they needed to improve the precision of their measurements. The French chemist Henri Victor Regnault was able to overcome many of these challenges and generate some of the most precise experimental data about the properties of gases at that time. Rudolf Clausius, a German physicist, then used Regnault's data to calculate the earliest published value for R. This value for R, however, was not very precise by current standards. Fortunately, there have been numerous advancements in the methods and tools that chemists use to measure the properties of gases, and the value of R has become increasingly precise over time. There is, however, always room for improvement. In this investigation, you will have an opportunity to follow in the footsteps of Clapeyron, Regnault, and Clausius by designing, conducting trials of, refining, and then evaluating a method that can be used to calculate a precise value for the ideal gas constant.

Your Task

Design a method that can be used to calculate an accurate value of R inside the lab by generating a specific number of moles of gas at room temperature and then measuring the pressure or volume of the gas. As part of this process you will need to test, evaluate, and then refine your method. Your method should allow you to produce a consistent and accurate value for R.

The guiding question of this investigation is, **How can a value of R for the ideal gas law be accurately determined inside the laboratory?**

LAB 15

Materials
You may use any of the following materials during your investigation:

Consumables	Equipment
• 6 M hydrochloric acid (HCl) solution • Magnesium (Mg) ribbon	• Side-arm Erlenmeyer flask with stopper (50 ml) • Pneumatic trough • Plastic or rubber tubing (50 cm long) • Electronic or triple beam balance • Graduated cylinders (one each 50 ml, 100 ml, 250 ml, and 500 ml) • Glass test tube • Utility clamp • Ring stand • Gas pressure sensor • Temperature probe • Sensor interface

Safety Precautions
Follow all normal lab safety rules. Hydrochloric acid is corrosive to eyes, skin, and other body tissues. Your teacher will explain relevant and important information about working with the chemicals associated with this investigation. In addition, take the following safety precautions:

- Wear indirectly vented chemical-splash goggles and chemical-resistant gloves and apron while in the laboratory.
- Handle all glassware with care.
- Wash your hands with soap and water before leaving the laboratory.

Investigation Proposal Required?   ☐ Yes   ☐ No

Getting Started
In this lab, you will react magnesium metal with hydrochloric acid to produce a sample of hydrogen gas. The hydrogen gas produced by this reaction behaves mostly like an ideal gas. The equation for this chemical reaction is

$$Mg + 2HCl \rightarrow MgCl_2 + H_2$$

The first step in your investigation is to design a method that will allow you to obtain the pressure, volume, temperature, and number of moles of a sample of hydrogen so you can use these data to calculate the gas constant (R). There are several approaches that you can use. One approach is to produce a specific amount of gas (in moles) and then measure the pressure of that gas while holding the volume and temperature constant. A second approach is to produce a specific amount of gas (in moles) and then measure its volume while holding pressure and temperature constant. It is important for you to consider how you might be able to measure the various properties of a sample of hydrogen gas using

The Ideal Gas Law

How Can a Value of R for the Ideal Gas Law Be Accurately Determined Inside the Laboratory?

the equipment available. As you design your method, you should also think about the following questions:

- What type of measurements or observations will you need to record during your investigation?
- How often will you collect data and when will you do it?
- How will you make sure that your data are of high quality (i.e., how will you reduce error)?
- How will you keep track of the data you collect and how will you organize it?
- How will you determine if there is a difference between the two methods?
- What type of calculations will you need to make?

The second step in your investigation is to test and refine your method. To do this, use your method to obtain information about a sample of hydrogen gas and then use this information to calculate a value for R. The value that you calculate will likely be rather imprecise due to flaws in your method or poor measurements. Use what you have learned from your initial test to refine your method. Once you have refined your method, you will need to test it again. You should continue this process of testing and refining until it functions as intended. Your method will therefore go through numerous iterations.

The last step in this investigation will be to conduct a formal evaluation of your method. As part of the evaluation of your method, you will need to determine if you can use it to produce a consistent and accurate value for R. You will therefore need to determine what data you need to collect and how you will analyze it as part of the evaluation to show that your method works.

Connections to Crosscutting Concepts, the Nature of Science, and the Nature of Scientific Inquiry

As you work through your investigation, be sure to think about

- the importance of identifying causal relationships in science,
- how scientists develop and use system models to understand complex phenomena,
- the difference between laws and theories, and
- the nature and role of experiments in science.

Initial Argument

Once your group has finished collecting and analyzing your data, you will need to develop an initial argument. Your argument must include a *claim*, which is your answer to the guiding question. Your argument must also include *evidence* in support of your claim. The

LAB 15

FIGURE L15.1
Argument presentation on a whiteboard

The Guiding Question:	
Our Claim:	
Our Evidence:	Our Justification of the Evidence:

evidence is your analysis of the data and your interpretation of what the analysis means. Finally, you must include a *justification* of the evidence in your argument. You will therefore need to use a scientific concept or principle to explain why the evidence that you decided to use is relevant and important. You will create your initial argument on a whiteboard. Your whiteboard must include all the information shown in Figure L15.1.

Argumentation Session

The argumentation session allows all of the groups to share their arguments. One member of each group stays at the lab station to share that group's argument, while the other members of the group go to the other lab stations one at a time to listen to and critique the arguments developed by their classmates. The goal of the argumentation session is not to convince others that your argument is the best one; rather, the goal is to identify errors or instances of faulty reasoning in the initial arguments so these mistakes can be fixed. You will therefore need to evaluate the content of the claim, the quality of the evidence used to support the claim, and the strength of the justification of the evidence included in each argument that you see. To critique an argument, you might need more information than what is included on the whiteboard. You might therefore need to ask the presenter one or more follow-up questions, such as:

- How did your group collect the data? Why did you use that method?
- What did your group do to make sure the data you collected are reliable? What did you do to decrease measurement error?
- What did your group do to analyze the data, and why did you decide to do it that way? Did you check your calculations?
- Is that the only way to interpret the results of your group's analysis? How do you know that your interpretation of the analysis is appropriate?
- Why did your group decide to present your evidence in that manner?
- What other claims did your group discuss before deciding on that one? Why did you abandon those alternative ideas?
- How confident are you that your group's claim is valid? What could you do to increase your confidence?

Once the argumentation session is complete, you will have a chance to meet with your group and revise your original argument. Your group might need to gather more data or design a way to test one or more alternative claims as part of this process. Remember, your goal at this stage of the investigation is to develop the most valid or acceptable answer to the research question!

The Ideal Gas Law
How Can a Value of R for the Ideal Gas Law Be Accurately Determined Inside the Laboratory?

Report

Once you have completed your research, you will need to prepare an *investigation report* that consists of three sections. Each section should provide an answer for the following questions:

1. What question were you trying to answer and why?
2. What did you do during your investigation and why did you conduct your investigation in this way?
3. What is your argument?

Your report should answer these questions in two pages or less. The report must be typed and any diagrams, figures, or tables should be embedded into the document. Be sure to write in a persuasive style; you are trying to convince others that your claim is acceptable or valid!

LAB 15

Checkout Questions

Lab 15. The Ideal Gas Law: How Can a Value of *R* for the Ideal Gas Law Be Accurately Determined Inside the Laboratory?

1. Describe the relationship between the pressure, volume, temperature, and number of moles of a gas.

2. One day when Jessica was at the grocery store she noticed two types of birthday balloons she could buy for her friend's upcoming party. One type of balloon was made out of latex, which is a type of rubber, and the other was made out of a substance called Mylar, which is like a very thin aluminum foil. Jessica bought one balloon of each type and had the store clerk fill them both with the same amount of helium. When she released the balloons in her house they both floated all the way to the ceiling, but the next morning only the Mylar balloon was still at the ceiling while the latex balloon was midway between the ceiling and the floor.

    Use what you know about the ideal gas law to explain what conditions must have been present for the balloons to behave in this manner.

The Ideal Gas Law
How Can a Value of R for the Ideal Gas Law Be Accurately Determined Inside the Laboratory?

3. Measuring the changes in volume in response to changes in temperature to test if they are directly related is an example of an experiment.

    a. I agree with this statement.

    b. I disagree with this statement.

    Explain your answer, using an example from your investigation about ideal gas law.

4. Laws are theories that have been developed and refined over time.

    a. I agree with this statement.

    b. I disagree with this statement.

    Explain your answer, using an example from your investigation about the ideal gas law.

5. An important goal in science is to develop causal explanations for observations. Explain what a causal explanation is and why these explanations are important, using an example from your investigation about the ideal gas law.

LAB 15

6. Scientists often use models to help them understand natural phenomena. Explain what a model is and why models are important, using an example from your investigation about the ideal gas law.

7. Theories can become laws over time.

    a. I agree with this statement.
    b. I disagree with this statement.

    Explain your answer, using an example from your investigation about bond character and molecular polarity.

8. Scientists often use models to help them understand natural phenomena. Explain what a model is and why models are important, using an example from your investigation about bond character and molecular polarity.

The Ideal Gas Law
How Can a Value of R for the Ideal Gas Law Be Accurately Determined Inside the Laboratory?

9. Scientists often look for and attempt to explain patterns in nature. Explain why patterns are important, using an example from your investigation about bond character and molecular polarity.

10. In nature, the structure of an object is often related to the function or properties of that object. Explain why this is true, using an example from your investigation about bond character and molecular polarity.

SECTION 3
Physical Sciences Core Idea 1.B

Chemical Reactions

Introduction Labs

LAB 16

Lab Handout

Lab 16. Development of a Reaction Matrix: What Are the Identities of the Unknown Chemicals?

Introduction

In chemistry it is often necessary to identify unknown substances. There are many procedures and pieces of equipment that can help a chemist to identify unknown substances, such as mass spectrometry or gas chromatography. While some of these methods and equipment are very complex and others are simpler, a common feature of each of these approaches is the comparison of properties of known substances with those of the unknown substances. Chemists can observe and compare *physical properties* of known and unknown substances, such as boiling point or density; likewise, it is possible to compare *chemical properties* such as how an unknown substance reacts with a known substance. One problem with testing and comparing physical properties of a substance involves solutions. It is possible that two different solutions will have the same density or even the same boiling point; in such a case those properties would not be helpful for differentiating between the two substances. In this situation, it may be more helpful to use observations of chemical reactions involving the unknown substance to determine what is in the solution.

An *aqueous solution* is created when a substance, the *solute*, is dissolved into water, the *solvent*. When different aqueous solutions are mixed together, it is possible that the substances dissolved into the water will undergo a chemical reaction, which may produce a *precipitate*. A precipitate is an insoluble ionic compound. Adding a few drops of an unknown solution to a powder may also cause a reaction, which might result in the formation of a gas. Other observable outcomes of a chemical reaction might be a change in color or a change in temperature. Since the chemical properties of a given substance are consistent when chemicals react with each other, the resulting products are also consistent and predictable; therefore, combining solutions together can be used as a way to identify them. This systematic process can be used to develop a table of reactions and their results, or a *reaction matrix*. If you know how specific known chemicals react with each other, it is possible to identify unknown chemicals by comparison.

Your Task

You will be given six labeled bottles to test and record data in order to generate a reaction matrix. Once your observations are complete and your reaction matrix has been developed, you will turn in your original set of chemicals. Then you will be given a second set of the same chemicals that are missing the correct labels. These unknowns will only be labeled as A–F. Your task will be to match the known solutions to the

Development of a Reaction Matrix
What Are the Identities of the Unknown Chemicals?

unknown solutions by comparing reactions with your written observations and the results in your reaction matrix.

The guiding question of this investigation is, **What are the identities of the unknown chemicals?**

Materials

You may use any of the following materials during your investigation:

Consumables	Equipment
• 6 known solutions • 6 unknown solutions (labeled A–F)	• Well plates or test tubes • Toothpicks • Dropper bottles or disposable pipettes

Safety Precautions

Follow all normal lab safety rules. Your teacher will explain relevant and important information about working with the chemicals associated with this investigation. Some of these chemicals may stain your skin and clothing. Take the following safety precautions in addition to any precautions specified by your teacher:

- Wear indirectly vented chemical-splash goggles and chemical-resistant gloves and apron while in the laboratory.
- Handle all glassware with care.
- Wash your hands with soap and water before leaving the laboratory.

Investigation Proposal Required?   ☐ Yes   ☐ No

Getting Started

To answer the guiding question, you will need to generate a reaction matrix. A reaction matrix is a specific style of data table that allows you to systematically document what happens when you react a series of chemicals with each other (see Table L16.1, p. 136). The reaction matrix accounts for all of the different possible reactions within the set of substances you have available.

TABLE L16.1

Example of a reaction matrix

Solution	Solution				
	A	B	C	D	E
A	Observation 1	Observation 2	Observation 3	Observation 4	Observation 5
B	Observation 6	Observation 7	Observation 8	Observation 9	Observation 10
C	Observation 11	Observation 12	Observation 13	Observation 14	Observation 15

The solutions being tested are listed as entries in the first column and as headers in the remaining columns. Observations made during each test can be recorded in the boxes.

To generate a reaction matrix, you must also determine what type of data you will need to collect, how you will collect the data, and how you will analyze the data.

To determine *what type of data you need to collect*, think about what types of observations or measurements will be most useful to include in your reaction matrix.

To determine *how you will collect the data*, think about the following questions:

- How much of each chemical will you need to mix together?
- How often will you collect data and when will you do it?
- How will you make sure that your data are of high quality (i.e., how will you reduce error)?
- How will you keep track of the data you collect and how will you organize it?

To determine *how you will analyze the data*, think about how you will determine if a known substance and an unknown substance match.

Connections to Crosscutting Concepts, the Nature of Science, and the Nature of Scientific Inquiry

As you work through your investigation, be sure to think about

- the importance of identifying patterns in science,
- the difference between observations and inferences in science, and
- the nature and role of experiments in science.

Initial Argument

Once your group has finished collecting and analyzing your data, you will need to develop an initial argument. Your argument must include a *claim*, which is your answer to the guiding question. Your argument must also include *evidence* in support of your claim. The

Development of a Reaction Matrix
What Are the Identities of the Unknown Chemicals?

evidence is your analysis of the data and your interpretation of what the analysis means. Finally, you must include a *justification* of the evidence in your argument. You will therefore need to use a scientific concept or principle to explain why the evidence that you decided to use is relevant and important. You will create your initial argument on a whiteboard. Your whiteboard must include all the information shown in Figure L16.1.

FIGURE L16.1
Argument presentation on a whiteboard

The Guiding Question:	
Our Claim:	
Our Evidence:	Our Justification of the Evidence:

Argumentation Session

The argumentation session allows all of the groups to share their arguments. One member of each group stays at the lab station to share that group's argument, while the other members of the group go to the other lab stations one at a time to listen to and critique the arguments developed by their classmates. The goal of the argumentation session is not to convince others that your argument is the best one; rather, the goal is to identify errors or instances of faulty reasoning in the initial arguments so these mistakes can be fixed. You will therefore need to evaluate the content of the claim, the quality of the evidence used to support the claim, and the strength of the justification of the evidence included in each argument that you see. To critique an argument, you might need more information than what is included on the whiteboard. You might, therefore, need to ask the presenter one or more follow-up questions, such as:

- What did your group do to analyze the data, and why did you decide to do it that way?
- Is that the only way to interpret the results of your group's analysis? How do you know that your interpretation of the analysis is appropriate?
- Why did your group decide to present your evidence in that manner?
- What other claims did your group discuss before deciding on that one? Why did you abandon those alternative ideas?
- How confident are you that your group's claim is valid? What could you do to increase your confidence?

Once the argumentation session is complete, you will have a chance to meet with your group and revise your original argument. Your group might need to gather more data or design a way to test one or more alternative claims as part of this process. Remember, your goal at this stage of the investigation is to develop the most valid or acceptable answer to the research question!

LAB 16

Report

Once you have completed your research, you will need to prepare an *investigation report* that consists of three sections that provide answers to the following questions:

1. What question were you trying to answer and why?
2. What did you do during your investigation and why did you conduct your investigation in this way?
3. What is your argument?

Your report should answer these questions in two pages or less. The report must be typed and any diagrams, figures, or tables should be embedded into the document. Be sure to write in a persuasive style; you are trying to convince others that your claim is acceptable or valid!

Checkout Questions

Lab 16. Development of a Reaction Matrix: What Are the Identities of the Unknown Chemicals?

1. Describe why chemical properties are useful for identifying matter.

2. Sometimes in chemistry two clear aqueous solutions, A and B, can be mixed and a reaction will occur. The result of the reaction is a new solid compound, C, that settles to the bottom of the container. Even though the initial solutions were clear, the new solid compound can have a very different color such as yellow, green, blue, or white.

   Use what you know about chemical reactions to explain where the matter for compound C comes from.

LAB 16

3. In science, observations are facts, whereas inferences are just guesses.

   a. I agree with this statement.
   b. I disagree with this statement.

   Explain your answer, using an example from your investigation about the development of a reaction matrix.

4. All investigations in chemistry are experiments.

   a. I agree with this statement.
   b. I disagree with this statement.

   Explain your answer, using an example from your investigation about the development of a reaction matrix.

5. Scientists conduct many investigations in hopes of identifying a pattern within nature. Using an example from your investigation about the development of a reaction matrix, describe how patterns can be useful for predicting the outcome of a new investigation.

LAB 17

Lab Handout

Lab 17. Limiting Reactants: Why Does Mixing Reactants in Different Mole Ratios Affect the Amount of the Product and the Amount of Each Reactant That Is Left Over?

Introduction

Atomic theory is a model that has been developed over time to explain the properties and behavior of matter. This model consists of five important principles, as listed below (see also Figure L17.1):

1. All matter is composed of submicroscopic particles called atoms.
2. All atoms of a given element are identical.
3. All atoms of one element have the same mass, and atoms from different elements have different masses.
4. Atoms can be combined with other atoms to form molecules, and molecules can be split apart into individual atoms.
5. Atoms are not created or destroyed during a chemical reaction that results in the production of a new substance.

A chemical reaction, according to atomic theory, is simply the rearrangement of atoms. The substances (elements and/or compounds) that are changed into other substances during a chemical reaction are called *reactants*. The substances that are produced as a result of a chemical reaction are called *products*. Chemical equations show the reactants and products of a chemical reaction. A chemical equation includes the chemical formulas of the reactants and the products. The products and reactants are separated by an arrow symbol ($\rightarrow$), and each individual substance's chemical formula is separated by a plus sign (+).

Atomic theory, as noted earlier, indicates that atoms are not created or destroyed during a chemical reaction. Thus, each side of the chemical equation must include the same number of each type of atom. When there is an equal number of each type of atom on each side of the equation, the equation is described as balanced. Chemists balance chemical equations by changing the number of each type of substance involved in the reaction; they do not change the number of atoms within each substance. The number of atoms found within a substance cannot be changed because it would change the nature of the substances involved in the reaction. For example, suppose a chemist needs to balance the following equation for the reaction of nitrogen and hydrogen gas:

Limiting Reactants

Why Does Mixing Reactants in Different Mole Ratios Affect the Amount of the Product and the Amount of Each Reactant That Is Left Over?

$$N_2(g) + H_2(g) \rightarrow NH_3(g)$$

In this case, he or she cannot simply add another atom of nitrogen and take an atom of hydrogen away from the chemical formula for ammonia (NH_3) because this would change ammonia (NH_3) to diazene (N_2H_2). Chemists therefore use stoichiometric coefficients to indicate how much of each substance is involved in the reaction without changing the nature of those substances. The balanced chemical equation for the reaction of nitrogen and hydrogen gas, as a result, is denoted as follows:

$$N_2(g) + 3\ H_2(g) \rightarrow 2\ NH_3(g)$$

FIGURE L17.1
The basic principles of atomic theory

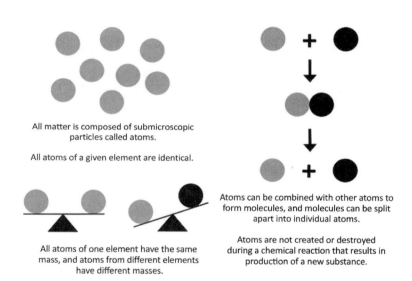

The stoichiometric coefficients indicate the relative amount of each substance involved in the chemical reaction in terms of molecules or moles. This equation, as a result, can be read as, 1 molecule of nitrogen gas reacts with 3 molecules of hydrogen gas to yield 2 molecules of ammonia gas. The equation can also be read as, 1 mole of nitrogen gas reacts with 3 moles of hydrogen gas to yield 2 moles of ammonia gas.

The stoichiometric coefficients are also used to determine the *mole ratio*, which is the relationship between the amounts of any two compounds, in moles, that are involved in a chemical reaction. For example, the mole ratio of the reactants (N_2 and H_2) in the reaction listed above is 1:3 (1 mole of N_2 reacts with 3 moles of H_2) and the mole ratio between hydrogen gas and ammonia gas is 3:2 (3 moles of H_2 yields 2 moles of NH_3). Chemists use mole ratios as a conversion factor in many chemistry problems. Mole ratios are also useful when

LAB 17

a chemist needs to create a specific amount of a product but must also minimize the amount of reactants that need to be purchased and limit the amount of waste that is left over.

To be able to use mole ratios to create a specific amount of a product, minimize costs, and limit waste, it is important to understand how varying the mole ratio of the reactants affects the amount of the product that is formed and the amount of the reactants that are left over at the end of a chemical reaction. You will therefore explore how mixing acetic acid and sodium bicarbonate in different mole ratios affects the amount of carbon dioxide (CO_2) gas that is produced and determine which reactant, if any, is left over at the end of the chemical reaction. You will then develop a conceptual model that you can use to explain your observations and predict the amount of CO_2 gas that will be produced in other conditions.

Your Task

Determine how varying the mole ratio of the reactants affects the amount of the product that is produced and the amount of the reactants that remain at the end of a chemical reaction. You will then develop a conceptual model that can be used to explain why mixing reactants in different mole ratios will affect the amount of product that is produced and the amount of each reactant left over. Once you have developed your conceptual model, you will need to test it to determine if it allows you to predict the dissolution rate of another solute under various conditions.

The guiding question for this investigation is, **Why does mixing reactants in different mole ratios affect the amount of the product and the amount of each reactant that is left over?**

Materials

You may use any of the following materials during your investigation:

Consumables	Equipment
• 1 M solution of acetic acid, CH_3COOH • Sodium bicarbonate, $NaHCO_3$ • pH paper	• Side-arm Erlenmeyer flask with stopper (50 ml) • Pneumatic trough • Tubing (50 cm long) • Electronic or triple beam balance • Graduated cylinder (500 ml) • Graduated cylinder (250 ml) • Spatula or chemical scoop • Weighing paper or dishes

Safety Precautions

Follow all normal lab safety rules. Your teacher will explain relevant and important information about working with the chemicals associated with this investigation. In addition, take the following safety precautions:

Limiting Reactants
Why Does Mixing Reactants in Different Mole Ratios Affect the Amount of the Product and the Amount of Each Reactant That Is Left Over?

- Wear indirectly vented chemical-splash goggles and chemical-resistant gloves and apron while in the laboratory.
- Handle all glassware with care.
- Wash your hands with soap and water before leaving the laboratory.

Investigation Proposal Required?   ☐ Yes   ☐ No

Getting Started

The first step in developing your model is to design and carry an experiment to determine how varying the mole ratio of the reactants affects the amount of the product that is formed and the amount of the reactants that remain at the end of a chemical reaction. To conduct this experiment, you will focus on the reaction of acetic acid and sodium bicarbonate. Acetic acid (CH_3COOH) reacts with sodium bicarbonate ($NaHCO_3$) according to the following equation:

$$CH_3COOH(aq) + NaHCO_3(s) \rightarrow NaCH_3CO_2(aq) + CO_2(g) + H_2O(l)$$

You will need to react sodium bicarbonate and acetic acid in different molar ratios while keeping everything else the same during your experiment. To do this, you will first need to decide which mole ratios (e.g., 2:1, 1:1, 1:3) and how many different mole ratios to test. You should, however, test at least five different mole ratios to have enough useful comparisons. Next, you will need to determine the amount of each reactant you need to use in each reaction. You should use the same amount of acetic acid (5 ml or 0.005 moles) in each reaction and vary the amount of sodium bicarbonate. You will therefore need to first determine the number of moles of sodium bicarbonate that you will need to use in each test. You can then calculate the mass of sodium bicarbonate that you will need to react with the 5 ml of acetic acid.

Once you have determined the amount of sodium bicarbonate you will need to use for each of the reactions, you will need to devise a way to determine how much product is produced during each reaction and if there are one or more reactants left over after the reaction is complete. You will therefore need to think about *what type of data you need to collect* during your experiment. To accomplish this task, think about the following questions:

- How will you know if there is any sodium bicarbonate left over at the end of the reaction?
- How will you know if there is any acetic acid left over at the end of the reaction?
- How will you measure the amount of product that is produced?

The easiest way to determine how much product is produced is to measure the amount of CO_2 that is formed after you combine the acetic acid and sodium bicarbonate. To accom-

LAB 17

FIGURE L17.2
Gas collection using water displacement

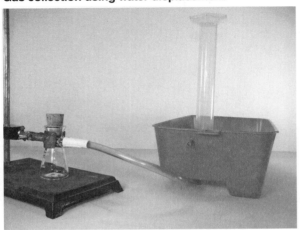

plish this task, you will need to collect the CO_2 gas by water displacement. Figure L17.2 shows how to collect gas by water displacement.

Once you have collected your data, you will need to think about *how you will analyze the data*. The following questions may be helpful:

- What types of comparisons will you make?
- What type of calculations will you need to make?
- What type of graph could you create to help you make sense of your data?

Once you have carried out your experiments, your group will need to develop a conceptual model that can be used to explain why mixing reactants in different mole ratios will affect the amount of product that is produced and the amount of each reactant left over. The model also needs to be able to explain what is happening during the reaction at the submicroscopic level.

The last step in this investigation is to test your model. To accomplish this goal, you can use a different mole ratio to determine if your model leads to accurate predictions about the amount of product produced and which reactant or reactants will be left over under different conditions. If you are able to use your model to make accurate predictions under different conditions, then you will be able to generate the evidence you need to convince others that the conceptual model you developed is valid.

Connections to Crosscutting Concepts, the Nature of Science, and the Nature of Scientific Inquiry

As you work through your investigation, be sure to think about

- why it is important to look for proportional relationships,
- how models are used to help understand natural phenomena,
- the difference between laws and theories in science, and
- the role of imagination and creativity in science.

Initial Argument

Once your group has finished collecting and analyzing your data, you will need to develop an initial argument. Your argument must include a *claim*, which is your answer to the guiding question. Your argument must also include *evidence* in support of your claim. The evidence is your analysis of the data and your interpretation of what the analysis means.

Limiting Reactants

Why Does Mixing Reactants in Different Mole Ratios Affect the Amount of the Product and the Amount of Each Reactant That Is Left Over?

Finally, you must include a *justification* of the evidence in your argument. You will therefore need to use a scientific concept or principle to explain why the evidence that you decided to use is relevant and important. You will create your initial argument on a whiteboard. Your whiteboard must include all the information shown in Figure L17.3.

FIGURE L17.3
Argument presentation on a whiteboard

The Guiding Question:	
Our Claim:	
Our Evidence:	Our Justification of the Evidence:

Argumentation Session

The argumentation session allows all of the groups to share their arguments. One member of each group stays at the lab station to share that group's argument, while the other members of the group go to the other lab stations one at a time to listen to and critique the arguments developed by their classmates. The goal of the argumentation session is not to convince others that your argument is the best one; rather, the goal is to identify errors or instances of faulty reasoning in the initial arguments so these mistakes can be fixed. You will therefore need to evaluate the content of the claim, the quality of the evidence used to support the claim, and the strength of the justification of the evidence included in each argument that you see. To critique an argument, you might need more information than what is included on the whiteboard. You might, therefore, need to ask the presenter one or more follow-up questions, such as:

- How did your group collect the data? Why did you use that method?
- What did your group do to make sure the data you collected are reliable? What did you do to decrease measurement error?
- What did your group do to analyze the data, and why did you decide to do it that way? Did you check your calculations?
- Is that the only way to interpret the results of your group's analysis? How do you know that your interpretation of the analysis is appropriate?
- Why did your group decide to present your evidence in that manner?
- What other claims did your group discuss before deciding on that one? Why did you abandon those alternative ideas?
- How confident are you that your group's claim is valid? What could you do to increase your confidence?

Once the argumentation session is complete, you will have a chance to meet with your group and revise your original argument. Your group might need to gather more data or design a way to test one or more alternative claims as part of this process. Remember, your goal at this stage of the investigation is to develop the most valid or acceptable answer to the research question!

LAB 17

Report

Once you have completed your research, you will need to prepare an *investigation report* that consists of three sections that provide answers to the following questions:

1. What question were you trying to answer and why?
2. What did you do during your investigation and why did you conduct your investigation in this way?
3. What is your argument?

Your report should answer these questions in two pages or less. The report must be typed and any diagrams, figures, or tables should be embedded into the document. Be sure to write in a persuasive style; you are trying to convince others that your claim is acceptable or valid!

Checkout Questions

Lab 17. Limiting Reactants: Why Does Mixing Reactants in Different Mole Ratios Affect the Amount of the Product and the Amount of Each Reactant That Is Left Over?

Use the following information to answer questions 1 and 2. Iron(III) chloride reacts with sodium hydroxide as follows:

$$FeCl_3(aq) + 3\,NaOH(aq) \rightarrow Fe(OH)_3(s) + 3\,NaCl(aq)$$

A student starts with 50 g of $FeCl_3$ and adds NaOH in 1-gram increments. He or she measures the mass of $Fe(OH)_3$ produced in each reaction and then plots the mass of $Fe(OH)_3$ as a function of the mass of NaOH added (see graph below).

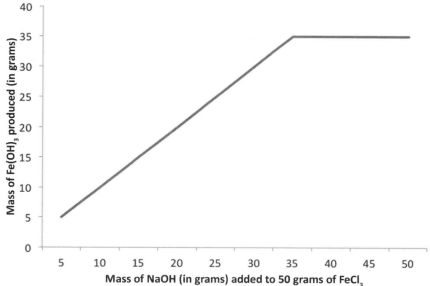

LAB 17

1. Describe the concept of a limiting reactant.

2. Use what you know about limiting reactants to explain why the mass of Fe(OH)$_3$ no longer changes when more than 37 g of NaOH is added to 50 g of FeCl$_3$.

3. Imagination and creativity play an important role in science.

    a. I agree with this statement.
    b. I disagree with this statement.

    Explain your answer, using an example from your investigation about limiting reactants.

Limiting Reactants
Why Does Mixing Reactants in Different Mole Ratios Affect the Amount of the Product and the Amount of Each Reactant That Is Left Over?

4. Scientists use laws to describe and theories to explain natural phenomena.

    a. I agree with this statement
    b. I disagree with this statement

    Explain your answer, using an example from your investigation about limiting reactants.

5. Scientists often need to look for proportional relationships. Explain why looking for a proportional relationship is useful in science, using an example from your investigation about limiting reactants.

6. Models are important in science. Explain what a model is and why models are important, using an example from your investigation about limiting reactants.

LAB 18

Lab Handout

Lab 18. Characteristics of Acids and Bases: How Can the Chemical Properties of an Aqueous Solution Be Used to Identify It as an Acid or a Base?

Introduction

Acids and bases represent two important classes of chemical compounds. These compounds play a significant role in many atmospheric and geological processes. In addition, acid-base reactions affect many of the physiological processes that take place within the human body. Acids and bases are important in atmospheric, geological, and physiological processes because they have unique chemical properties. Acids and bases have unique chemical properties because of the atomic composition of these compounds and how these compounds interact with other atoms and molecules.

Some of the unique chemical properties of acids and bases include how they interact with metals, carbonates, and a class of compounds called acid-base indicators. A *metal* is a solid material that is hard, shiny, malleable, and ductile. Metals are also good electrical and thermal conductors. *Carbonates* are compounds that contain a carbonate ion (CO_3^{2-}), such as calcium carbonate ($CaCO_3$), potassium carbonate (K_2CO_3), and sodium bicarbonate ($NaHCO_3$). An acid-base indicator is a dye or a pigment that changes color when it is mixed with an acid or a base. People have used indicators to identify acids and bases for hundreds of years. For example, in the 17th century Sir Robert Boyle described how different indicators could be used to identify acids and bases (Boyle 1664).

In this investigation you will explore some of the unique chemical properties of acids and bases. You will then develop a method that can be used to identify acidic or basic aqueous solutions. This is important because, like all chemists, you will need to be able to determine if an aqueous solution is acidic or basic as part of your future investigations. This is an important aspect of doing acid-base chemistry.

Your Task

Devise, test, and then, if needed, refine a method that can be used to determine if an aqueous solution is acidic or a basic. For this method to be useful, it should provide consistent and accurate results but should also be simple and quick to perform inside the lab.

The guiding question for this investigation is, **How can the chemical properties of an aqueous solution be used to identify it as an acid or a base?**

Characteristics of Acids and Bases
How Can the Chemical Properties of an Aqueous Solution Be Used to Identify It as an Acid or a Base?

Materials

You may use any of the following materials during this investigation:

Consumables	Indicators	Equipment	Aqueous solutions for developing a method	Aqueous solutions for testing a method
• Zinc • 1 M solution of $NaHCO_3$ • 1 M solution of hydrochloric acid, HCl	• Thymol blue • Bromphenol blue • Bromthymol blue • Methyl red • Phenol red	• Conductivity tester or probe • Reaction plate • Small beakers	• Acid solution 1 • Acid solution 2 • Acid solution 3 • Acid solution 4 • Base solution 1 • Base solution 2 • Base solution 3 • Base solution 4	• Acid test solution A • Acid test solution B • Base test solution A • Base test solution B

Safety Precautions

Follow all normal lab safety rules. All of the acids you will use are corrosive to eyes, skin, and other body tissues. They are also toxic when ingested. Your teacher will explain relevant and important information about working with the chemicals associated with this investigation. In addition, take the following safety precautions:

- Wear indirectly vented chemical-splash goggles and chemical-resistant gloves and apron while in the laboratory.
- Handle all glassware with care.
- Wash your hands with soap and water before leaving the laboratory.

Investigation Proposal Required?   ☐ Yes   ☐ No

Getting Started

To answer the guiding question, you will first need to learn more about the unique chemical properties of acids and bases. You will therefore need to explore how aqueous solutions that are classified as acids or as bases react with metal, a solution of sodium bicarbonate, and a solution of hydrochloric acid. You will then determine if these same solutions are able to conduct electricity. Finally, and perhaps most important, you will examine how different acidic and basic solutions interact with different indicators. You goal is to learn more about the chemical properties of aqueous solutions that are classified as being acids or bases so you can use these unique properties to classify other aqueous solutions. To accomplish this task, you will need to design and conduct a series of systematic observations.

Be sure to think about *how you will collect your data and how you will analyze the data you collect* before you begin your investigation. One way to collect data is to add a small amount (about 5 to 10 drops) of each acid or base solution to the wells in a reaction plate.

LAB 18

You can then add a small piece of metal or other solution to each well and observe what happens. You can also create a reaction matrix to help stay organized. A reaction matrix is a chart that allows you to record your observations (see Table L18.1 for an example). Only use the solutions found under the heading "Aqueous Solutions for Developing a Method" in the "Materials" section during this stage of your investigation.

TABLE L18.1
Example of a reaction matrix

Compound	Test				
	Zinc	Conductivity	HCl	Bromthymol blue	Methyl red
Acid solution 1	Observation 1	Observation 2	Observation 3	Observation 4	Observation 5
Acid solution 2	Observation 6	Observation 7	Observation 8	Observation 9	Observation 10
Base solution 1	Observation 11	Observation 12	Observation 13	Observation 14	Observation 15

Notice that the compounds being tested are included in the first column and each test is labeled as a header in the remaining columns. Observations made during each test can be recorded in the boxes.

Once you have made your observations about the chemical properties of acids and bases, you will need to use what you have learned to devise a method for classifying an unknown as either an acid or a base. You can then test your method using the solutions found under the heading "Aqueous Solutions for Testing a Method" in the "Materials" section. If you are able to use your method to accurately classify all four of these solutions, then you will be able to provide evidence that the method you devised will provide accurate results. If you cannot accurately classify all four of the test solutions, you will need to refine your method and test it again. Keep in mind that your method needs to be a simple and quick way to classify an unknown aqueous solution based on its chemical properties.

Connections to Crosscutting Concepts, the Nature of Science, and the Nature of Scientific Inquiry

As you work through your investigation, be sure to think about

- the importance of looking for, using, and explaining patterns in science;
- the relationship between structure and function in nature;
- the difference between observations and inferences in science; and
- the different methods used in scientific investigations.

Characteristics of Acids and Bases
How Can the Chemical Properties of an Aqueous Solution Be Used to Identify It as an Acid or a Base?

Initial Argument

Once your group has finished collecting and analyzing your data, you will need to develop an initial argument. Your argument must include a *claim*, which is your answer to the guiding question. Your argument must also include *evidence* in support of your claim. The evidence is your analysis of the data and your interpretation of what the analysis means. Finally, you must include a *justification* of the evidence in your argument. You will therefore need to use a scientific concept or principle to explain why the evidence that you decided to use is relevant and important. You will create your initial argument on a whiteboard. Your whiteboard must include all the information shown in Figure L18.1.

Argumentation Session

The argumentation session allows all of the groups to share their arguments. One member of each group stays at the lab station to share that group's argument, while the other members of the group go to the other lab stations one at a time to listen to and critique the arguments developed by their classmates. The goal of the argumentation session is not to convince others that your argument is the best one; rather, the goal is to identify errors or instances of faulty reasoning in the initial arguments so these mistakes can be fixed. You will therefore need to evaluate the content of the claim, the quality of the evidence used to support the claim, and the strength of the justification of the evidence included in each argument that you see. To critique an argument, you might need more information than what is included on the whiteboard. You might, therefore, need to ask the presenter one or more follow-up questions, such as:

FIGURE L18.1
Argument presentation on a whiteboard

The Guiding Question:	
Our Claim:	
Our Evidence:	Our Justification of the Evidence:

- How did your group collect the data? Why did you use that method?
- What did your group do to make sure the data you collected are reliable?
- What did your group do to analyze the data, and why did you decide to do it that way?
- Is that the only way to interpret the results of your group's analysis?
- Why did your group decide to present your evidence in that manner?
- What other claims did your group discuss before deciding on that one? Why did you abandon those alternative ideas?
- How confident are you that your group's claim is valid? What could you do to increase your confidence?

Once the argumentation session is complete, you will have a chance to meet with your group and revise your original argument. Your group might need to gather more data or design a way to test one or more alternative claims as part of this process. Remember, your goal at this stage of the investigation is to develop the most valid or acceptable answer to the research question!

Report

Once you have completed your research, you will need to prepare an *investigation report* that consists of three sections that provide answers to the following questions:

1. What question were you trying to answer and why?
2. What did you do during your investigation and why did you conduct your investigation in this way?
3. What is your argument?

Your report should answer these questions in two pages or less. The report must be typed and any diagrams, figures, or tables should be embedded into the document. Be sure to write in a persuasive style; you are trying to convince others that your claim is acceptable or valid!

Reference

Boyle, R. 1664. *Experiments and considerations touching colours first occasionally written, among some other essays to a friend, and now suffer'd to come abroad as the beginning of an experimental history of colours*. London: Henry Herringman.

Checkout Questions

Lab 18. Characteristics of Acids and Bases: How Can the Chemical Properties of an Aqueous Solution Be Used to Identify It as an Acid or a Base?

1. Describe three characteristics of acids and three characteristics of bases.

2. An unknown solution conducts electricity but the indicators thymol blue and bromphenol blue do not change color when they are added to it. Should this solution be classified as an acid or a base?

    a. Acid
    b. Base
    c. Not enough information to determine

    Explain your answer.

3. "The solution is an acid" is an example of an observation.

    a. I agree with this statement.
    b. I disagree with this statement.

    Explain your answer, using an example from your investigation about the characteristics of acids and bases.

LAB 18

4. An investigation must follow the scientific method to be considered scientific.

    a. I agree with this statement.
    b. I disagree with this statement.

    Explain your answer, using an example from your investigation about the characteristics of acids and bases.

5. Scientists often look for and attempt to explain patterns in nature. Explain why patterns are important, using an example from your investigation about the characteristics of acids and bases.

6. In nature, the structure of an object is often related to function or the properties of that object. Explain why this is true, using an example from your investigation about the characteristics of acids and bases.

Strong and Weak Acids
Why Do Strong and Weak Acids Behave in Different Manners Even Though They Have the Same Chemical Properties?

Lab Handout

Lab 19. Strong and Weak Acids: Why Do Strong and Weak Acids Behave in Different Manners Even Though They Have the Same Chemical Properties?

Introduction

Johannes Nicolaus Brønsted and Thomas Martin Lowry published nearly identical explanations for the nature of acids and bases in 1923. These two explanations were later combined into a single explanation, which is now known as the Brønsted-Lowry acid-base theory. This theory defines acids and bases in terms of how molecules interact with *hydrogen ions*. A hydrogen ion is just a proton (see Figure L19.1). An *acid*, according to the Brønsted-Lowry definition, is any substance from which a proton can be removed, and a base is any substance that can remove a proton from an acid molecule. In an acid, the hydrogen ion is bonded to the rest of the molecule. It therefore takes energy to break that bond. So an acid molecule does not "give up" or "donate" a proton, it has it taken away. When a base molecule interacts with an acid molecule, it will (if it is strong enough) rip the proton off the acid molecule.

FIGURE L19.1
A hydrogen atom and a hydrogen ion

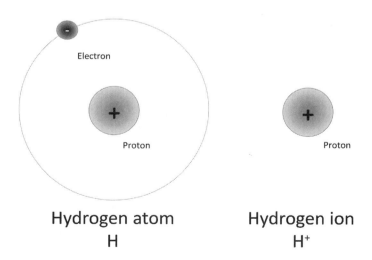

To illustrate how this works, consider what happens when hydrogen chloride (HCl) is mixed with water. In this situation, the water is able to remove a proton from the hydrogen chloride. The hydrogen chloride is therefore an acid and the water is therefore a base. In Brønsted's original explanation for the nature of an acid-base interaction, he used H^+ to describe how the proton is removed from an acid by a base. This interaction between the molecules of hydrogen chloride and water can therefore be represented as

$$HCl + H_2O \leftarrow \rightarrow H^+ + Cl^- + H_2O$$

A hydrogen ion (H^+), however, does not exist for very long in water because the proton affinity of H_2O is approximately 799 kJ/mol. As a result, a hydrogen ion quickly combines with a water molecule once it is removed from an acid molecule. This can be represented as

$$H_2O + H^+ \rightarrow H_3O^+$$

Lowry therefore used H_3O^+ rather than H^+ to describe the transfer of protons in his explanation of what happens on a submicroscopic level during an acid-base reaction. In his explanation, Lowry explained that when an acid is added to water, a proton from the acid is split off and taken up by water (the base) to produce hydronium ions (H_3O^+). For the hydrogen chloride and water example, this can be represented as

$$HCl + H_2O \leftarrow \rightarrow Cl^- + H_3O^+$$

The Brønsted-Lowry acid-base theory was a groundbreaking idea when it was first introduced because it was able to explain a wide range of macroscopic observations about the behavior of acids and bases by providing a model for how acid and base molecules interact with each other on the submicroscopic level. In this investigation, you will use the Brønsted-Lowry definition for an acid and a base as a starting point to develop a model that can be used to explain the behavior of strong and weak acids.

Your Task

Develop a model that can be used to explain why strong and weak acids behave in a different manner even though they have the same chemical properties. The two chemical properties of acids that you will focus on during this investigation are electrical conductivity and reactivity with metal. To develop your model, you will first need to determine how acid strength affects electrical conductivity and reaction rate. You will then need to determine how to explain your observations on the macroscopic level by describing the nature or behavior of strong and weak acids on the submicroscopic level. The Brønsted-Lowry definition of acids and bases will serve as the theoretical foundation for your model.

The guiding question for this investigation is, **Why do strong and weak acids behave in different manners even though they have the same chemical properties?**

Strong and Weak Acids

Why Do Strong and Weak Acids Behave in Different Manners Even Though They Have the Same Chemical Properties?

Materials

You may use any of the following materials during this investigation:

Consumables	Equipment
• 1 M solution of acetic acid, CH_3COOH (weak acid) • 1 M solution of hydrochloric acid, HCl (strong acid) • 1 M solution of sulfuric acid, H_2SO_4 (strong acid) • Magnesium ribbon	• Conductivity tester or probe • Pneumatic trough and tubing • Graduated cylinder (250 ml) • Graduated cylinder (10 ml) • Filtering flask (50 ml) and rubber stopper • Stopwatch • Electronic or triple beam balance

Safety Precautions

Follow all normal lab safety rules. All of the acids you will use are corrosive to eyes, skin, and other body tissues. They are also toxic by ingestion. Magnesium metal is a flammable solid and burns with an intense flame. Keep away from flames. Your teacher will explain relevant and important information about working with the chemicals associated with this investigation. In addition, take the following safety precautions:

- Wear indirectly vented chemical-splash goggles and chemical-resistant gloves and apron while in the laboratory.
- Handle all glassware with care.
- Wash your hands with soap and water before leaving the laboratory.

Investigation Proposal Required?   ☐ Yes   ☐ No

Getting Started

To answer the guiding question, you will need to first determine how the behavior of strong and weak acids differ in terms of electrical conductivity and reactivity with metals. A conductivity tester or probe can be used to measure the conductivity of the three different acid solutions. You can design and carry out an experiment to determine how acid strength affects reactivity with metal. All of the acids that you will be using react with magnesium to produce hydrogen gas. Your goal is to determine the relationship between acid strength and the rate of this reaction. You will therefore need to determine what type of data to collect, how you will collect the data, and how you will analyze the data to accomplish your goal.

To determine *what type of data you need to collect*, think about the following questions:

- What type of measurements will you need to make during your investigation? You could, for example, measure the amount of H_2 gas that is produced, the time it takes to produce a set amount of gas, or the time it takes for the reaction to go to completion.

LAB 19

- When will you need to take your measurements?

To determine *how you will collect the data*, think about the following questions:

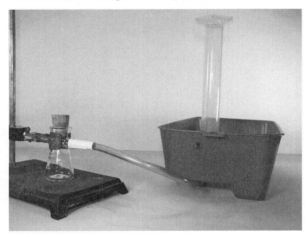

FIGURE L19.2

Gas collection using water displacement

- What equipment can you use to capture and measure the volume of a gas? You could, for example, capture and measure the volume of a gas using water displacement (see Figure L19.2).
- What types of test conditions will you need to set up and how will you do it?
- How will you eliminate confounding variables?
- How will you make sure that your data are of high quality (i.e., how will you reduce error)?
- How often will you collect data and when will you do it?
- How will you keep track of the data you collect and how will you organize it?

To determine how you will analyze the data, think about the following questions:

- How will you determine if there is a difference between the test conditions?
- What type of calculations will you need to make?
- What type of graph could you create to help make sense of your data?

Once you have carried out your experiment, your group will need to develop your conceptual model. The model should be able to explain why strong and weak acids differ in terms of electrical conductivity and reactivity with metal. The model should also include a description of the interactions that take place between molecules and should be based on the Brønsted-Lowry definition of acids and bases.

Connections to Crosscutting Concepts, the Nature of Science, and the Nature of Scientific Inquiry

As you work through your investigation, be sure to think about

- the importance of developing causal explanations for observations,
- how models are used to help understand natural phenomena,
- the importance of imagination and creativity in science, and
- the nature and role of experiments in science.

Strong and Weak Acids

Why Do Strong and Weak Acids Behave in Different Manners Even Though They Have the Same Chemical Properties?

Initial Argument

Once your group has finished collecting and analyzing your data, you will need to develop an initial argument. Your argument must include a *claim*, which is your answer to the guiding question. Your argument must also include *evidence* in support of your claim. The evidence is your analysis of the data and your interpretation of what the analysis means. Finally, you must include a *justification* of the evidence in your argument. You will therefore need to use a scientific concept or principle to explain why the evidence that you decided to use is relevant and important. You will create your initial argument on a whiteboard. Your whiteboard must include all the information shown in Figure L19.3.

FIGURE L19.3
Argument presentation on a whiteboard

The Guiding Question:	
Our Claim:	
Our Evidence:	Our Justification of the Evidence:

Argumentation Session

The argumentation session allows all of the groups to share their arguments. One member of each group stays at the lab station to share that group's argument, while the other members of the group go to the other lab stations one at a time to listen to and critique the arguments developed by their classmates. The goal of the argumentation session is not to convince others that your argument is the best one; rather, the goal is to identify errors or instances of faulty reasoning in the initial arguments so these mistakes can be fixed. You will therefore need to evaluate the content of the claim, the quality of the evidence used to support the claim, and the strength of the justification of the evidence included in each argument that you see. To critique an argument, you might need more information than what is included on the whiteboard. You might, therefore, need to ask the presenter one or more follow-up questions, such as:

- How did your group collect the data? Why did you use that method?
- What did your group do to make sure the data you collected are reliable? What did you do to decrease measurement error?
- What did your group do to analyze the data, and why did you decide to do it that way? Did you check your calculations?
- Is that the only way to interpret the results of your group's analysis? How do you know that your interpretation of the analysis is appropriate?
- Why did your group decide to present your evidence in that manner?
- What other claims did your group discuss before deciding on that one? Why did you abandon those alternative ideas?
- How confident are you that your group's claim is valid? What could you do to increase your confidence?

LAB 19

Once the argumentation session is complete, you will have a chance to meet with your group and revise your original argument. Your group might need to gather more data or design a way to test one or more alternative claims as part of this process. Remember, your goal at this stage of the investigation is to develop the most valid or acceptable answer to the research question!

Report

Once you have completed your research, you will need to prepare an *investigation report* that consists of three sections that provide answers to the following questions:

1. What question were you trying to answer and why?
2. What did you do during your investigation and why did you conduct your investigation in this way?
3. What is your argument?

Your report should answer these questions in two pages or less. The report must be typed and any diagrams, figures, or tables should be embedded into the document. Be sure to write in a persuasive style; you are trying to convince others that your claim is acceptable or valid!

Checkout Questions

Lab 19. Strong and Weak Acids: Why Do Strong and Weak Acids Behave in Different Manners Even Though They Have the Same Chemical Properties?

1. Acids and bases are useful reactants in the chemistry laboratory and play an important role in biology and nature. According to the Brønsted-Lowry acid-base theory, what is an acid and what is a base?

2. Strong acids react with reactive metals at a faster rate and conduct electricity better than weak acids. Use what you know about the characteristics of strong and weak acids to explain why the strong acids have a faster reaction rate and are better at conducting electricity than weak acids.

3. Measuring the electrical conductivity of a solution is an example of an experiment.

    a. I agree with this statement.
    b. I disagree with this statement.

   Explain your answer, using an example from your investigation about strong and weak acids.

LAB 19

4. Scientists do not need to be creative or have a good imagination to excel in science.

    a. I agree with this statement.
    b. I disagree with this statement.

    Explain your answer, using an example from your investigation about strong and weak acids.

5. An important goal in science is to develop causal explanations for observations. Explain what a causal explanation is and why these explanations are important, using an example from your investigation about strong and weak acids.

6. Scientists often use models to help them understand natural phenomena. Explain what a model is and why models are important, using an example from your investigation about strong and weak acids.

Enthalpy Change of Solution
How Can Chemists Use the Properties of a Solute to Predict If an Enthalpy Change of Solution Will Be Exothermic or Endothermic?

Lab Handout

Lab 20. Enthalpy Change of Solution: How Can Chemists Use the Properties of a Solute to Predict If an Enthalpy Change of Solution Will Be Exothermic or Endothermic?

Introduction

Thermodynamics is the study of energy changes in a system. Thermodynamics is an important field of study in chemistry because energy changes occur during chemical reactions, when solutes are dissolved in solvents, and when matter goes through a change of state. Chemists often describe the energy changes that take place in these situations in terms of heat content. *Enthalpy* is a measure of the heat content of a system. The transfer of heat into or out of a system results in a change in enthalpy. This change in the heat content of a system is symbolized as ΔH (delta H). The unit of measurement for an enthalpy change is kilojoules per mole (kJ/mol).

The enthalpy change that occurs when a solute is dissolved in water is called the heat of solution or the *enthalpy change of solution* ($\Delta H_{solution}$). The enthalpy change of solution is equal in magnitude to the heat energy lost from or gained by the surroundings. When heat energy is lost from the system and gained by the surroundings, the enthalpy change of solution is described as *exothermic*. An *endothermic* enthalpy change of solution, in contrast, occurs when the system gains heat energy from the surroundings.

The overall energy change that occurs when a solute is dissolved in water (i.e., the $\Delta H_{solution}$) is the result of two key processes. First, an input of energy breaks the attractive forces holding the particles in the solute together and disrupts the hydrogen bonds holding the water molecules together. The system *gains* energy and the surroundings *lose* energy during this process. For the purposes of this investigation, this change in energy will be called the *enthalpy change of dissociation* ($\Delta H_{dissociation}$). Second, energy is released as attractive forces form between the particles of the solute and the molecules of water. The system *loses* energy and the surroundings *gain* energy during this process. The energy change that occurs during this process is called the *enthalpy change of hydration* ($\Delta H_{hydration}$). An illustration of the energy inputs and outputs associated with dissolution of an ionic compound is provided in Figure L20.1 (p. 168).

As described earlier, the dissolution of a solute involves both a gain and a loss of energy, and the $\Delta H_{solution}$ of a solute can be either endothermic or exothermic depending on the net amount of energy that is lost from or gained by the system. The net energy change of the system will depend, in part, on the unique properties of the solute. Solutes can be composed

LAB 20

FIGURE L20.1
The dissociation process and hydration process that take place when an ionic compound dissolves in water

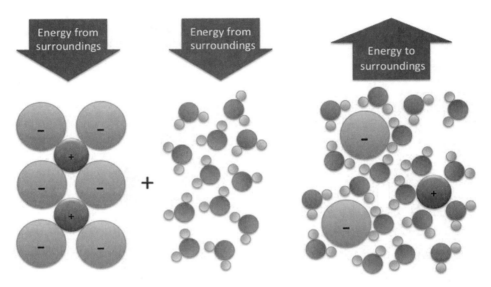

Enthalpy change of dissociation
Energy is absorbed by the system
Ions separate and water molecules separate

Enthalpy change of hydration
Energy is released by the system
Ions and water molecules combine

of different types of particles, and attractive forces hold these particles together. The nature and strength of these attractive forces will influence the amount of energy that is required to break apart the solute. In addition, the particles that make up a solute will differ in terms of the strength of their electrical charge. Some solute particles, as a result, will attract water molecules better than others. The strength of attraction that exists between solute particles and water molecules will influence the amount of energy that is released when the solute particles and the water molecules combine. The physical and chemical properties of the solute, as a result, will affect the $\Delta H_{solution}$.

In this investigation, you will be given five different ionic compounds. You will then determine if the $\Delta H_{solution}$ for each ionic compound is endothermic or exothermic by mixing the solute with water and measuring the resulting temperature change. Next, you will use a table of $\Delta H_{dissociation}$ and $\Delta H_{hydration}$ values to develop a rule that you can use to predict if the $\Delta H_{solution}$ of other ionic compounds will be endothermic or exothermic. The $\Delta H_{dissociation}$ values reflect the amount of energy needed to separate the ions in the solute and the amount of energy needed to disrupt the attractive forces between the water molecules. The $\Delta H_{hydration}$ values reflect the energy that is released when ion-dipole

Enthalpy Change of Solution
How Can Chemists Use the Properties of a Solute to Predict If an Enthalpy Change of Solution Will Be Exothermic or Endothermic?

forces form between the individual ions and the water molecules. You will then be given an opportunity to test your rule with two other ionic compounds to determine if you can use it to make accurate predictions.

Your Task

Develop a rule that chemists can use to determine if the enthalpy change of solution ($\Delta H_{solution}$) for a given ionic compound will be endothermic or exothermic based on the properties of the solute.

The guiding question for this investigation is, **How can chemists use the properties of a solute to predict if an enthalpy change of solution will be exothermic or endothermic?**

Materials

You may use any of the following materials during this investigation:

Consumables	Equipment
• Calcium chloride, $CaCl_2$, 5 grams • Cesium chloride, CsCl, 5 grams • Lithium chloride, LiCl, 5 grams • Potassium chloride, KCl, 5 grams • Sodium chlorate, $NaClO_3$, 5 grams • Sodium chloride, NaCl, 5 grams • Sodium iodide, NaI, 5 grams • Distilled water	• 2 polystyrene cups (or a calorimeter) • Thermometer (or temperature probe and sensor interface) • Graduated cylinder (25 ml) • 3 beakers (each 250 ml) • Stirring rod • Electronic or triple beam balance • Timer or stopwatch • Support stand and ring clamp • Chemical scoop • Weighing paper or dishes

Safety Precautions

Follow all normal lab safety rules. Lithium chloride, calcium chloride, cesium chloride, sodium chlorate, and sodium iodide are all moderately toxic by ingestion and are tissue irritants. Your teacher will explain relevant and important information about working with the chemicals associated with this investigation. In addition, take the following safety precautions:

- Wear indirectly vented chemical-splash goggles and chemical-resistant gloves and apron while in the laboratory.
- Handle all glassware with care.
- Wash your hands with soap and water before leaving the laboratory.

LAB 20

Investigation Proposal Required? ☐ Yes   ☐ No

Getting Started

The first step in developing your rule is to determine if the $\Delta H_{solution}$ for LiCl, $CaCl_2$, KCl, NaCl, and CsCl is endothermic or exothermic. To accomplish this step, you will need to dissolve each ionic compound in water and measure the resulting temperature change using a calorimeter. A calorimeter is an insulated container that is designed to prevent heat loss to the atmosphere. A simple calorimeter can be made from two polystyrene cups, a support stand, and a ring clamp (see Figure L20.2). Once you have set up a simple calorimeter, you must determine what type of data you need to collect, how you will collect the data, and how you will analyze the data.

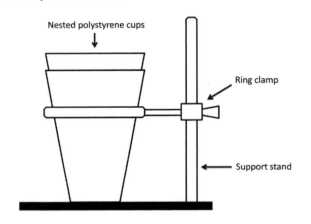

FIGURE L20.2
A simple calorimeter

To determine *what type of data you need to collect*, think about the following questions:

- What type of measurements or observations will you need to record during each test?
- How often will you need to make these measurements or observations?

To determine *how you will collect the data*, think about the following questions:

- How much water will you use in the calorimeter?
- Will the same amount of water be used for each test?
- How much of each ionic compound will you need to use?
- Will you need to use the same amount of each ionic compound for each test? If so, does it need to be the same amount in terms of mass or in terms of moles?
- What will you do to reduce measurement error?
- How will you keep track of the data you collect and how will you organize it?

To determine *how you will analyze the data*, think about the following questions:

- What type of calculations will you need to make?
- What type of graph could you create to help make sense of your data?

Once you have carried out your series of tests, your group will need to develop your rule. Table L20.1 provides the $\Delta H_{dissociation}$ and $\Delta H_{hydration}$ values for each ionic compound. Positive values indicate that the energy is being absorbed by the system, and negative

Enthalpy Change of Solution
How Can Chemists Use the Properties of a Solute to Predict If an Enthalpy Change of Solution Will Be Exothermic or Endothermic?

values indicate that energy is being released by the system. As you develop your rule, think about how you could present it as a mathematical equation.

TABLE L20.1

$\Delta H_{dissociation}$ and $\Delta H_{hydration}$ values for the ionic compounds used in this investigation

Ionic compound	$\Delta H_{dissociation}$ (kJ/mol)*	$\Delta H_{hydration}$ (kJ/mol)
LiCl	853	−883
CaCl$_2$	2,258	−2378
KCl	715	−685
NaCl	788	−784
CsCl	657	−639
NaClO$_3$	658	−636
NaI	693	−701

* This value reflects the energy required to break apart the ions in the solute and the energy required to disrupt the hydrogen bonds between the water molecules.

The last step is to test your rule. To accomplish this goal, you will need to determine if you can use your rule to make accurate predictions. You should, in other words, be able to use your rule to predict if the $\Delta H_{solution}$ of NaClO$_3$ and NaI will be exothermic or endothermic. If you are able to make accurate predictions about the $\Delta H_{solution}$ for these two ionic compounds, then you will be able to generate the evidence you need to convince others that the rule that you developed is valid.

Connections to Crosscutting Concepts, the Nature of Science, and the Nature of Scientific Inquiry

As you work through your investigation, be sure to think about

- the importance of defining the system under study and then using a model to make sense of it,
- the importance of tracking how matter and energy move into and within a system,
- the difference between observations and inferences in science, and
- the difference between laws and theories in science.

Initial Argument

Once your group has finished collecting and analyzing your data, you will need to develop an initial argument. Your argument must include a *claim*, which is your answer to the guiding question. Your argument must also include *evidence* in support of your

LAB 20

FIGURE L20.3
Argument presentation on a whiteboard

The Guiding Question:	
Our Claim:	
Our Evidence:	Our Justification of the Evidence:

claim. The evidence is your analysis of the data and your interpretation of what the analysis means. Finally, you must include a *justification* of the evidence in your argument. You will therefore need to use a scientific concept or principle to explain why the evidence that you decided to use is relevant and important. You will create your initial argument on a whiteboard. Your whiteboard must include all the information shown in Figure L20.3.

Argumentation Session

The argumentation session allows all of the groups to share their arguments. One member of each group stays at the lab station to share that group's argument, while the other members of the group go to the other lab stations one at a time to listen to and critique the arguments developed by their classmates. The goal of the argumentation session is not to convince others that your argument is the best one; rather, the goal is to identify errors or instances of faulty reasoning in the initial arguments so these mistakes can be fixed. You will therefore need to evaluate the content of the claim, the quality of the evidence used to support the claim, and the strength of the justification of the evidence included in each argument that you see. To critique an argument, you might need more information than what is included on the whiteboard. You might, therefore, need to ask the presenter one or more follow-up questions, such as:

- What did your group do to make sure the data you collected are reliable? What did you do to decrease measurement error?
- What did your group do to analyze the data, and why did you decide to do it that way? Did you check your calculations?
- Is that the only way to interpret the results of your group's analysis? How do you know that your interpretation of the analysis is appropriate?
- Why did your group decide to present your evidence in that manner?
- What other claims did your group discuss before deciding on that one? Why did you abandon those alternative ideas?
- How confident are you that your group's claim is valid? What could you do to increase your confidence?

Once the argumentation session is complete, you will have a chance to meet with your group and revise your original argument. Your group might need to gather more data or design a way to test one or more alternative claims as part of this process. Remember, your goal at this stage of the investigation is to develop the most valid or acceptable answer to the research question!

Enthalpy Change of Solution
How Can Chemists Use the Properties of a Solute to Predict If an Enthalpy Change of Solution Will Be Exothermic or Endothermic?

Report

Once you have completed your research, you will need to prepare an *investigation report* that consists of three sections that provide answers to the following questions:

1. What question were you trying to answer and why?
2. What did you do during your investigation and why did you conduct your investigation in this way?
3. What is your argument?

Your report should answer these questions in two pages or less. The report must be typed and any diagrams, figures, or tables should be embedded into the document. Be sure to write in a persuasive style; you are trying to convince others that your claim is acceptable or valid!

Checkout Questions

Lab 20. Enthalpy Change of Solution: How Can Chemists Use the Properties of a Solute to Predict If an Enthalpy Change of Solution Will Be Exothermic or Endothermic?

Use the following information to answer questions 1–3. When chromium(II) chloride, $CrCl_2$, is dissolved in water, the temperature of the water decreases.

1. Is this an endothermic or exothermic process?

    a. Endothermic
    b. Exothermic

Explain why.

2. What is the $\Delta H_{solution}$?

    a. > 0
    b. < 0

Explain why.

Enthalpy Change of Solution
How Can Chemists Use the Properties of a Solute to Predict If an Enthalpy Change of Solution Will Be Exothermic or Endothermic?

3. Which is stronger, the attractive force between the water molecules and the chromium and chloride ions or the combined ionic bond strength of the $CrCl_2$ and the intermolecular forces between the water molecules?

    a. The attractive forces between the water molecules and chromium and chloride ions

    b. The combined ionic bond strength of the $CrCl_2$ and the intermolecular forces between the water molecules

   Use what you know about the process of dissolving to explain your answer.

4. "Heat flowed into the system" is an observation.

    a. I agree with this statement.
    b. I disagree with this statement.

   Explain your answer, using an example from your investigation about enthalpy change of solution.

LAB 20

5. Theories can turn into laws.

   a. I agree with this statement.
   b. I disagree with this statement.

   Explain your answer, using an example from your investigation about enthalpy change of solution.

6. Scientists often need to track how matter and energy move into, out of, and within a system. Explain why this is important, using an example from your investigation about enthalpy change of solution.

7. Scientists often need to define a system under study in order to study it. Explain why it is important to define a system under study, using an example from your investigation about enthalpy change of solution.

Lab Handout

Lab 21. Reaction Rates: Why Do Changes in Temperature and Reactant Concentration Affect the Rate of a Reaction?

Introduction

The molecular-kinetic theory of matter suggests that all matter is made up of submicroscopic particles called atoms that are constantly in motion. These atoms can be joined together to form molecules. Atoms have kinetic energy because they move and vibrate. The more kinetic energy an atom has, the faster it moves or vibrates. *Temperature* is a measurement of the average kinetic energy of all the atoms in a substance. The average kinetic energy of the particles within a substance increases and decreases as it changes temperature. *Heat*, in contrast, is the total kinetic energy of all the particles in a substance.

A chemical reaction, as you have learned, is simply the rearrangement of atoms. The substances (elements and/or compounds) that are changed into other substances during a chemical reaction are called *reactants*. The substances that are produced as a result of a chemical reaction are called *products*. Chemical equations show the reactants and products of a chemical reaction. A chemical equation includes the chemical formulas of the reactants and the products. The products and reactants are separated by an arrow symbol (→), and the chemical formula for each individual substance is separated by a plus sign (+).

A balanced chemical equation tells us the nature of the products and the amount of product that is formed from a given amount of reactants. A balanced chemical equation, however, tells us little about how long it takes for the reaction to happen. Some chemical reactions, such as the rusting of iron, happen slowly over time, while others, such as the burning of gasoline, are almost instantaneous. The speed of any reaction is indicated by its *reaction rate*, which is a measure of how quickly the reactants transform into products. As shown in Figure L21.1 (p. 178), a reaction begins with only reactant molecules. Over time, the reactant molecules interact with each other and form product molecules. The concentration of reactant molecules and the product molecules, as a result, will change during the process of a reaction. The rate of a reaction can therefore be calculated by measuring how the concentration of the reactants decreases or the concentration of the products increases as a function of time. The rate of a reaction can also be measured by timing how long it takes for a product to appear or for a reactant to disappear once the reaction begins.

LAB 21

FIGURE L21.1
A model of what happens during a chemical reaction over time

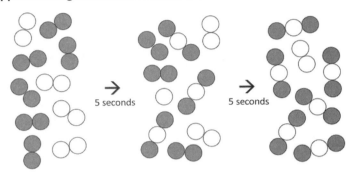

It is important for chemists to understand how and why different factors affect the rate of a chemical reaction so they can make a wide range of products in a safe and economical manner. You will therefore explore two factors that affect the rate of a specific reaction and then develop a conceptual model that you can use to explain your observations and predict the rate of this reaction under different conditions.

Your Task

Determine how temperature and changes in the concentration of a reactant affect the rate of the reaction between magnesium (Mg) and hydrochloric acid (HCl). Then develop a conceptual model that can be used to explain *why* these factors influence reaction rate. Once you have developed your model, you will need to determine if it is consistent with the rates of reaction that you observe under other conditions.

The guiding question of this investigation is, **Why do changes in temperature and reactant concentration affect the rate of a reaction?**

Materials

You may use any of the following materials during your investigation:

Consumables	Equipment
• 2 pieces of 18-gauge copper wire (20 cm) • Magnesium ribbon (30 cm) • 3 M HCl • 2 M HCl • 1 M HCl • 0.5 M HCl • 0.1 M HCl • Ice	• 2 Pyrex test tubes • Test tube rack • Graduated cylinder (25 ml) • 2 beakers (each 250 ml) • Thermometer (or temperature probe) • Hot plate • Electronic or triple beam balance • pH paper

Reaction Rates
Why Do Changes in Temperature and Reactant Concentration Affect the Rate of a Reaction?

Safety Precautions
Follow all normal lab safety rules. Magnesium is a flammable solid, and hydrochloric acid is a corrosive liquid. Your teacher will explain relevant and important information about working with the chemicals associated with this investigation. In addition, take the following safety precautions:

- Wear indirectly vented chemical-splash goggles and chemical-resistant gloves and apron while in the laboratory.
- Do not heat hydrochloric acid directly on a hot plate; rather, heat the hydrochloric acid using a hot bath and keep the temperature of the bath between 20°C and 60°C.
- Use caution when working with hot plates because they can burn skin. Hot plates also need to be kept away from water and other liquids.
- Handle all glassware with care.
- Wash your hands with soap and water before leaving the laboratory.

Investigation Proposal Required?   ☐ Yes   ☐ No

Getting Started
The first step in developing your model is to design and carry out two experiments. The goal of the first experiment will be to determine how temperature affects reaction rate. The goal of the second experiment will be to determine how reactant concentration affects reaction rate. For these two experiments, you will focus on the reaction of magnesium and hydrochloric acid. These two chemicals react to form hydrogen gas and magnesium chloride. The equation for this reaction is

$$Mg(s) + 2HCl(aq) \rightarrow H_2(g) + MgCl_2(aq)$$

You can measure the reaction rate by simply timing how long it takes for the solid magnesium to be no longer visible once it is mixed with the hydrochloric acid. You can also measure the reaction rate by timing how long hydrogen gas is produced after the magnesium and hydrochloric acid are mixed. The unit of measurement for a reaction rate is mol/sec. It will therefore be important for you to determine how many moles of each reactant you used for each test.

To design your two experiments, you will need to decide what type of data you need to collect, how you will collect the data, and how you will analyze the data.

To determine *what type of data you need to collect*, think about the following questions:

- How will you determine the number of moles of each reactant at the beginning of the reaction?

LAB 21

- How will you know when the reaction starts and when it is finished?
- What type of measurements will you need to record during each experiment?
- When will you need to make these measurements or observations?

To determine *how you will collect the data*, think about the following questions:

- How much magnesium ribbon will you use in each test?
- How much hydrochloric acid will you need to use to submerge the magnesium ribbon?
- How will you prevent the magnesium ribbon from floating on top of the hydrochloric acid? One way to prevent the magnesium from floating on top of the hydrochloric acid is to use copper wire to create a cage (see Figure L21.2).
- What will serve as your independent variable?
- What types of comparisons will you need to make?
- How will you hold other variables constant?
- What will you do to reduce measurement error?
- How will you keep track of the data you collect and how will you organize it?

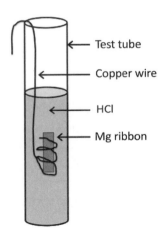

FIGURE L21.2

How to suspend magnesium (Mg) ribbon in the hydrochloric acid (HCl). The copper wire does not react with the HCl.

To determine *how you will analyze the data*, think about the following questions:

- What type of calculations will you need to make?
- What type of graph could you create to help make sense of your data?

Once you have carried out your two experiments, your group will need to develop a conceptual model. This conceptual model must provide an underlying reason for your findings about the effect of temperature and reactant concentration on reaction rate. Your model should also include an explanation of what is happening at the submicroscopic level between and within molecules during a reaction. The molecular-kinetic theory of matter should serve as the theoretical foundation for your model.

The last step in this investigation is to test your model. To accomplish this goal, you can use the same reaction but test different temperatures and concentrations to determine if your model is consistent with the rates of reactions you observe under different conditions. If you can use your model to make accurate predictions about the rate of this reaction under different conditions, then you will be able to generate the evidence you need to convince others that the conceptual model you developed is valid.

Reaction Rates
Why Do Changes in Temperature and Reactant Concentration Affect the Rate of a Reaction?

Connections to Crosscutting Concepts, the Nature of Science, and the Nature of Scientific Inquiry

As you work through your investigation, be sure to think about

- the importance of developing causal explanations for observations,
- how models are used to help understand natural phenomena,
- the importance of imagination and creativity in science, and
- the role of experiments in science.

Initial Argument

Once your group has finished collecting and analyzing your data, you will need to develop an initial argument. Your argument must include a *claim*, which is your answer to the guiding question. Your argument must also include *evidence* in support of your claim. The evidence is your analysis of the data and your interpretation of what the analysis means. Finally, you must include a *justification* of the evidence in your argument. You will therefore need to use a scientific concept or principle to explain why the evidence that you decided to use is relevant and important. You will create your initial argument on a whiteboard. Your whiteboard must include all the information shown in Figure L21.3.

FIGURE L21.3
Argument presentation on a whiteboard

The Guiding Question:	
Our Claim:	
Our Evidence:	Our Justification of the Evidence:

Argumentation Session

The argumentation session allows all of the groups to share their arguments. One member of each group stays at the lab station to share that group's argument, while the other members of the group go to the other lab stations one at a time to listen to and critique the arguments developed by their classmates. The goal of the argumentation session is not to convince others that your argument is the best one; rather, the goal is to identify errors or instances of faulty reasoning in the initial arguments so these mistakes can be fixed. You will therefore need to evaluate the content of the claim, the quality of the evidence used to support the claim, and the strength of the justification of the evidence included in each argument that you see. To critique an argument, you might need more information than what is included on the whiteboard. You might, therefore, need to ask the presenter one or more follow-up questions, such as:

- How did your group collect the data? Why did you use that method?
- What did your group do to make sure the data you collected are reliable? What did you do to decrease measurement error?

LAB 21

- What did your group do to analyze the data, and why did you decide to do it that way? Did you check your calculations?
- Is that the only way to interpret the results of your group's analysis? How do you know that your interpretation of the analysis is appropriate?
- Why did your group decide to present your evidence in that manner?
- What other claims did your group discuss before deciding on that one? Why did you abandon those alternative ideas?
- How confident are you that your group's claim is valid? What could you do to increase your confidence?

Once the argumentation session is complete, you will have a chance to meet with your group and revise your original argument. Your group might need to gather more data or design a way to test one or more alternative claims as part of this process. Remember, your goal at this stage of the investigation is to develop the most valid or acceptable answer to the research question!

Report

Once you have completed your research, you will need to prepare an *investigation report* that consists of three sections that provide answers to the following questions:

1. What question were you trying to answer and why?
2. What did you do during your investigation and why did you conduct your investigation in this way?
3. What is your argument?

Your report should answer these questions in two pages or less. The report must be typed and any diagrams, figures, or tables should be embedded into the document. Be sure to write in a persuasive style; you are trying to convince others that your claim is acceptable or valid!

Checkout Questions

Lab 21. Reaction Rates: Why Do Changes in Temperature and Reactant Concentration Affect the Rate of a Reaction?

Chemists must be able to measure and control the rate of a chemical reaction in order to produce substances in a safe and economical way. Chemists can slow down a reaction rate by lowering the temperature of the reaction or by diluting the concentration of the reactants.

1. Describe the concept of a reaction rate.

2. Describe the molecular-kinetic theory of matter.

3. Use what you know about reaction rates and the molecular-kinetic theory of matter to explain why lowering the temperature of a reaction or diluting the concentration of the reactants in a reaction will decrease the rate of a chemical reaction.

LAB 21

4. Scientists use experiments to prove ideas right or wrong.

    a. I agree with this statement.
    b. I disagree with this statement.

   Explain your answer, using an example from your investigation about reaction rates.

5. Scientists need to be creative and have a good imagination to excel in science.

    a. I agree with this statement.
    b. I disagree with this statement.

   Explain your answer, using an example from your investigation about reaction rates.

6. An important goal in science is to develop causal explanations for observations. Explain what a causal explanation is and why these explanations are important, using an example from your investigation about reaction rates.

7. Scientists often use or develop new models to help them understand natural phenomena. Explain what a model is in science and why models are important, using an example from your investigation about reaction rates.

LAB 22

Lab Handout

Lab 22. Chemical Equilibrium: Why Do Changes in Temperature, Reactant Concentration, and Product Concentration Affect the Equilibrium Point of a Reaction?

Introduction

It is often useful to think of a reaction as a process that consists of two components acting in opposite directions. From this view, a reaction begins with all reactants and no products. The reactants then begin to interact with each other and transform into products. The rate at which the reactants transform into products will begin to decrease over time as the concentration of the reactant decreases. At this point, some of the products will begin to revert back into reactants. The rate at which the products revert back into reactants will increase as the concentration of the product increases. There is a point, as a result, where the forward and reverse components of a reaction are happening at equal rates. This point is called *chemical equilibrium*. At equilibrium, the rates of the forward and reverse components of the reaction are equal but the concentrations of reactants and products are not. Figure L22.1 illustrates this process.

Chemical equilibrium, therefore, can be defined as the point in a reaction where the rate at which reactants transform into products is equal to the rate at which products revert back into reactants. The *equilibrium point* of a chemical reaction occurs when the amount or concentration of the products and reactants in a closed system is stable. Chemists use a specific property, such as color, concentration, or density, to determine when a reaction is in equilibrium. It is important to note, however, that chemists view the state of chemical equilibrium as dynamic because reactants continue to transform into products and products continue to revert back into reactants even though the amount of reactants and products in the closed system is stable.

The equilibrium point of a reaction can change because chemical equilibrium is not static. There are a number of different factors that can change the equilibrium point of a reaction by changing the rate at which reactants transform into products or by changing the rate at which products revert back into the reactants. These factors include a change in temperature, pressure, reactant concentration, and product concentration. When any of these factors are changed, the equilibrium point of the reaction will move and the concentration of products and reactants in the system at the new equilibrium point will be different.

Chemical Equilibrium
Why Do Changes in Temperature, Reactant Concentration, and Product Concentration Affect the Equilibrium Point of a Reaction?

FIGURE L22.1

The forward and reverse reactions associated with chemical equilibrium

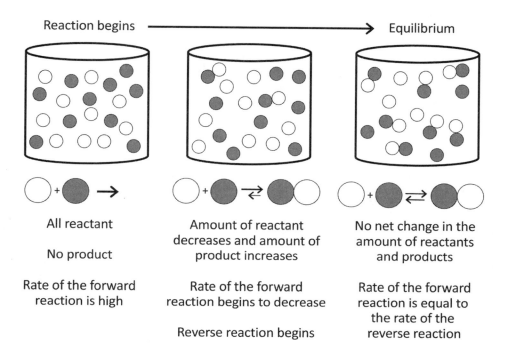

To control the amount of product or reactant present at the equilibrium point of a reaction in a closed system, chemists need to understand how various factors affect chemical equilibrium and why these various factors change the equilibrium point of a reaction. You will therefore explore how three specific factors affect the equilibrium point of chemical reaction. You will then develop a conceptual model that you can use to explain your observations and predict how the equilibrium point of a different reaction will change when the equilibrium point is disturbed by changing these same three factors.

Your Task

Determine *how* changes in temperature and the addition of extra reactant and product affect the equilibrium point of the reaction between iron(III) nitrate and potassium thiocyanate. Then develop a conceptual model that you can use to explain *why* these factors influence the equilibrium point of a reaction. Once you have developed your conceptual model, you will need to test it to determine if it allows you to predict how the equilibrium point of a different reaction will change under similar conditions.

The guiding question of this investigation is, **Why do changes in temperature, reactant concentration, and product concentration affect the equilibrium point of a reaction?**

LAB 22

Materials

You may use any of the following materials during your investigation:

Consumables	Equipment
• 0.1 M iron(III) nitrate, $Fe(NO_3)_3$ • 0.1 M potassium thiocyanate, KSCN • 1.5 M copper(II) chloride, $CuCl_2$ • 4 M sodium chloride, NaCl • 0.05 M silver nitrate, $AgNO_3$ • Distilled water • Ice	• 9 test tubes • Test tube rack • Graduated cylinder (10 ml) • 6 disposable graduated Beral pipettes • Beaker (50 ml) • 2 beakers (each 250 ml), for hot- and cold-water baths • Thermometer • Hot plate

Safety Precautions

Follow all normal lab safety rules. Iron(III) nitrate is a body tissue irritant; it will also stain clothes and skin. Potassium thiocyanate and copper(II) chloride are toxic by ingestion. Silver nitrate is toxic by ingestion, is corrosive to body tissues, and stains clothes and skin. Your teacher will explain relevant and important information about working with the chemicals associated with this investigation. In addition, take the following safety precautions:

- Wear indirectly vented chemical-splash goggles and chemical-resistant gloves and apron while in the laboratory.
- Use caution when working with hot plates because they can burn skin. Hot plates also need to be kept away from water and other liquids.
- Handle all glassware (including thermometers) with care.
- Wash your hands with soap and water before leaving the laboratory.

Investigation Proposal Required?   ☐ Yes   ☐ No

Getting Started

The first step in developing your model is to design and carry out three experiments. The goal of the first experiment will be to determine how a change in reactant concentration affects the equilibrium point of a reaction. The goal of the second experiment will be to determine how a change in product concentration affects the equilibrium point of a reaction. The goal of the third experiment will be to determine how temperature affects the equilibrium point of a reaction. For these three experiments, you will focus on the reaction of iron(III) nitrate and potassium thiocyanate. Iron(III) ions react with thiocyanate ions to form $FeSCN^{2+}$ complex ions according to the following reaction:

$$Fe^{3+}(aq) + SCN^-(aq) \leftrightarrows FeSCN^{2+}(aq)$$

Yellow   Colorless   Orange-Red

Chemical Equilibrium
Why Do Changes in Temperature, Reactant Concentration, and Product Concentration Affect the Equilibrium Point of a Reaction?

You can prepare a stock solution of $FeSCN^{2+}$ by mixing 40 ml of distilled water with 1 ml of 0.1 M $Fe(NO_3)_3$ and 2 ml of 0.1 M KSCN. You can then add 2 ml of this stock solution to several different test tubes to create a control condition and several treatment conditions for each experiment. You can then change the temperature, reactant concentration, or product concentration in the treatment conditions as needed and leave the control condition alone for comparison purposes. You must, however, determine what type of data you need to collect, how you will collect the data, and how you will analyze the data for each experiment.

To determine *what type of data you need to collect*, think about the following questions:

- What type of measurements or observations will you need to record during each experiment?
- When will you need to make these measurements or observations?

To determine *how you will collect the data*, think about the following questions:

- What will serve as your independent variable in each experiment?
- How will you change the independent variable in each experiment?
- What types of comparisons will you need to make?
- What will you do to reduce measurement error?
- How will you keep track of the data you collect and how will you organize it?

To determine *how you will analyze the data*, think about the following questions:

- What type of calculations will you need to make?
- What type of graph could you create to help make sense of your data?

Once you have carried out your three experiments, your group will need to develop a conceptual model. This conceptual model will need to be able to provide an underlying reason for your findings about the effect of temperature, changes in reactant concentration, and changes in product concentration on the equilibrium point of a reaction. Your model should also include an explanation of what is happening at the submicroscopic level between and within molecules during a reaction. The collision theory of reaction rates and the concept of chemical equilibrium should serve as the theoretical foundation for your model.

The last step in this investigation is to test your model. To accomplish this goal, you can use a different reaction to determine if your model leads to accurate predictions about how the equilibrium point changes in response to different factors. If you can use your model to make accurate predictions about how the equilibrium point of a different reversible reaction changes, then you will be able to generate the evidence you need to convince others that the conceptual model you developed is valid.

You can use the reversible formation of copper(II) complexes to test your model. When copper(II) chloride ($CuCl_2$) is dissolved in water, two different solutes are present in the solution. These solutes include Cu^{2+} ions and Cl^- ions. These solutes interact with water molecules to form two different complex ions. One complex ion is $Cu(H_2O)_6^{2+}$ and the other is $CuCl_4^{2-}$. The reversible equation for the formation of the two complex ions is

$$Cu(H_2O)_6^{2+}(aq) + 4Cl^-(aq) \leftrightarrows CuCl_4^{2-}(aq) + 6H_2O$$

Blue Green

You can change the equilibrium point by adding NaCl, or $AgNO_3$ or by changing the temperature of the solution. To change the concentration of the reactants or the products, simply add 2 ml of the copper(II) chloride solution to a test tube and then add up to eight drops of NaCl or $AgNO_3$. The addition of NaCl will increase the number of Cl^- ions in the system. The addition of $AgNO_3$, in contrast, will decrease the number of Cl^- ions in the system (because the Ag^+ ions react with Cl^- ions to form AgCl). To change the temperature of the system, use a hot-water bath or an ice bath.

Connections to Crosscutting Concepts, the Nature of Science, and the Nature of Scientific Inquiry

As you work through your investigation, be sure to think about

- how models are used to help understand natural phenomena,
- why it is important to understand what makes a system stable or unstable and what controls rates of change in a system,
- the importance of imagination and creativity in science, and
- the role of experiments in science.

Initial Argument

Once your group has finished collecting and analyzing your data, you will need to develop an initial argument. Your argument must include a *claim*, which is your answer to the guiding question. Your argument must also include *evidence* in support of your claim. The evidence is your analysis of the data and your interpretation of what the analysis means. Finally, you must include a *justification* of the evidence in your argument. You will therefore need to use a scientific concept or principle to explain why the evidence that you decided to use is relevant and important. You will create your initial argument on a whiteboard. Your whiteboard must include all the information shown in Figure L22.2.

Argumentation Session

The argumentation session allows all of the groups to share their arguments. One member of each group stays at the lab station to share that group's argument, while the other members

Chemical Equilibrium
Why Do Changes in Temperature, Reactant Concentration, and Product Concentration Affect the Equilibrium Point of a Reaction?

of the group go to the other lab stations one at a time to listen to and critique the arguments developed by their classmates. The goal of the argumentation session is not to convince others that your argument is the best one; rather, the goal is to identify errors or instances of faulty reasoning in the initial arguments so these mistakes can be fixed. You will therefore need to evaluate the content of the claim, the quality of the evidence used to support the claim, and the strength of the justification of the evidence included in each argument that you see. To critique an argument, you might need more information than what is included on the whiteboard. You might, therefore, need to ask the presenter one or more follow-up questions, such as:

FIGURE L22.2

Argument presentation on a whiteboard

The Guiding Question:	
Our Claim:	
Our Evidence:	Our Justification of the Evidence:

- How did your group collect the data? Why did you use that method?
- What did your group do to make sure the data you collected are reliable? What did you do to decrease measurement error?
- What did your group do to analyze the data, and why did you decide to do it that way? Did you check your calculations?
- Is that the only way to interpret the results of your group's analysis? How do you know that your interpretation of the analysis is appropriate?
- Why did your group decide to present your evidence in that manner?
- What other claims did your group discuss before deciding on that one? Why did you abandon those alternative ideas?
- How confident are you that your group's claim is valid? What could you do to increase your confidence?

Once the argumentation session is complete, you will have a chance to meet with your group and revise your original argument. Your group might need to gather more data or design a way to test one or more alternative claims as part of this process. Remember, your goal at this stage of the investigation is to develop the most valid or acceptable answer to the research question!

Report

Once you have completed your research, you will need to prepare an *investigation report* that consists of three sections that provide answers to the following questions:

1. What question were you trying to answer and why?
2. What did you do during your investigation and why did you conduct your investigation in this way?

LAB 22

3. What is your argument?

Your report should answer these questions in two pages or less. The report must be typed and any diagrams, figures, or tables should be embedded into the document. Be sure to write in a persuasive style; you are trying to convince others that your claim is acceptable or valid!

Checkout Questions

Lab 22. Chemical Equilibrium: Why Do Changes in Temperature, Reactant Concentration, and Product Concentration Affect the Equilibrium Point of a Reaction?

Chemists must understand chemical equilibrium in order to control the amount of product produced as a result of a reaction. Chemists can alter the equilibrium point of a reaction by changing the temperature or the concentration of the reactants or products.

1. What happens to the concentration of the reactants and the products in a system when a reaction is at equilibrium?

2. How is the rate at which the reactants transform into products related to the rate at which the products revert back into reactants when a reaction is at equilibrium?

3. Paper coated with cobalt chloride is sold commercially as test strips for estimating humidity. The following reversible reaction takes place between cobalt chloride and water:

$$CoCl_2(s) + H_2O(g) \rightleftharpoons CoCl_2 \cdot 6H_2O(s)$$

          Blue   Pink

   a. What color will the test strip be when the humidity is low (20%), and what color will it be when the humidity is high (80%)? Explain your answer.

   b. The test strips come with a color chart to estimate intermediate humidity levels. What color do you think the paper will be when the humidity is about 50%? Explain your answer.

4. Scientists use experiments to test potential explanations for a phenomenon.

   a. I agree with this statement.

   b. I disagree with this statement.

Explain your answer, using an example from your investigation about chemical equilibrium.

Chemical Equilibrium
Why Do Changes in Temperature, Reactant Concentration, and Product Concentration Affect the Equilibrium Point of a Reaction?

5. Scientists do not need to be creative or have a good imagination.

    a. I agree with this statement.
    b. I disagree with this statement.

    Explain your answer, using an example from your investigation about chemical equilibrium.

6. An important goal in science is to understand what makes a system stable or unstable and what controls rates of change in system. Explain why this important, using an example from your investigation about chemical equilibrium.

7. Scientists often use or develop new models to help them understand natural phenomena. Explain what a model is in science and why models are important, using an example from your investigation about chemical equilibrium.

Application Labs

LAB 23

Lab Handout

Lab 23. Classification of Changes in Matter: Which Changes Are Examples of a Chemical Change, and Which Are Examples of a Physical Change?

Introduction

Matter, the "stuff" of which the universe is composed, has two characteristics: it has mass and it occupies space. Physical properties of matter, such as density, odor, color, melting point, boiling point, state at room temperature (liquid, gas or solid), and magnetism, are often useful for identifying different substances. Matter, however, can also go through changes in both its physical and chemical properties. During *physical changes* the composition of matter does not change; for example, freezing a sample of water results in a change of state (i.e., going from liquid to solid), but the substance is still water (H_2O)—its chemical composition did not change. During *chemical changes* the chemical composition of a substance does change; for example, burning a piece of wood in a fireplace is a chemical change. In this example, the original wood is transformed into ashes and smoke, which both have different chemical properties than the original piece of wood.

Your Task

Create and observe the five scenarios listed below. Using your data and observations, determine if a physical or chemical change has occurred when

1. 100 ml of water (H_2O) is mixed with 5 g of table salt (NaCl),
2. a 2 cm magnesium strip is placed in a crucible and heated,
3. 10 drops of sodium hydroxide (NaOH) and 10 drops of copper(II) nitrate ($CuNO_3$) are mixed,
4. 5 drops of hydrochloric acid (HCl) are added to 2 g of sodium bicarbonate ($NaHCO_3$), and
5. paraffin wax is subjected to heat in a hot-water bath.

The guiding question of this investigation is, **Which changes are examples of a chemical change, and which are examples of a physical change?**

Classification of Changes in Matter
Which Changes Are Examples of a Chemical Change, and Which Are Examples of a Physical Change?

Materials

You may use any of the following materials during your investigation:

Consumables	Equipment
• NaCl	• Spot (reaction) plate
• NaOH solution	• Graduated cylinder (50 ml)
• $CuNO_3$ solution	• Beaker (150 ml)
• HCl solution	• Beaker (500 ml)
• $NaHCO_3$	• Hot plate
• Magnesium strip, 2 cm	• Bunsen burner
• Paraffin wax	• Ring stand with metal ring
• pH paper	• Clay triangle
• Distilled water (in squirt bottles)	• Wire gauze square
	• Crucible with lid
	• Crucible tongs
	• Test tube tongs
	• Spatula
	• Thermometer
	• Electronic or triple beam balance

Safety Precautions

Follow all normal lab safety rules. Your teacher will explain relevant and important information about working with the chemicals associated with this investigation. In addition, take the following safety precautions:

- Wear indirectly vented chemical-splash goggles and chemical-resistant gloves and apron while in the laboratory.
- Wipe up any water spilled on the floor from water baths.
- When investigating the odor associated with chemicals, never inhale with your nose directly over a tube, beaker or bottle; your instructor will demonstrate wafting the fumes toward your nose with your hand.
- Use caution when working with Bunsen burners. They can burn skin, and combustibles and flammables must be kept away from the open flame. If you have long hair, tie it back behind your head.
- Inspect the crucible for cracks. If it is cracked, exchange it for a new one. Clean the crucible and lid thoroughly before using them.
- Be careful with a crucible after removing it from a flame because it will still be hot.
- Use caution when working with hot plates, hot water, and melted wax because they can burn skin. Hot plates also need to be kept away from water and other liquids.
- Handle test tubes placed in the hot-water bath ONLY with test tube tongs.
- Handle all glassware (including thermometers) with care.
- Wash your hands with soap and water before leaving the laboratory.

LAB 23

Investigation Proposal Required?   ☐ Yes   ☐ No

Getting Started

Create each of the scenarios listed on the previous page and record what happens. Then conduct additional tests as needed to determine if a chemical or physical change took place.

Connections to Crosscutting Concepts, the Nature of Science, and the Nature of Scientific Inquiry

As you work through your investigation, be sure to think about

- the importance of patterns within science,
- the flow of energy and matter within a system,
- the difference between observations and inferences in science, and
- the difference between data and evidence in science.

Initial Argument

Once your group has finished collecting and analyzing your data, you will need to develop an initial argument. Your argument must include a *claim*, which is your answer to the guiding question. Your argument must also include *evidence* in support of your claim. The evidence is your analysis of the data and your interpretation of what the analysis means. Finally, you must include a *justification* of the evidence in your argument. You will therefore need to use a scientific concept or principle to explain why the evidence that you decided to use is relevant and important. You will create your initial argument on a whiteboard. Your whiteboard must include all the information shown in Figure L23.1.

FIGURE L23.1
Argument presentation on a whiteboard

The Guiding Question:	
Our Claim:	
Our Evidence:	Our Justification of the Evidence:

Argumentation Session

The argumentation session allows all of the groups to share their arguments. One member of each group stays at the lab station to share that group's argument, while the other members of the group go to the other lab stations one at a time to listen to and critique the arguments developed by their classmates. The goal of the argumentation session is not to convince others that your argument is the best one; rather, the goal is to identify errors or instances of faulty reasoning in the initial arguments so these mistakes can be fixed. You will therefore need to evaluate the content of the claim, the quality of the evidence used to support the claim, and the strength of the justification of the evidence included in each argument that you see. To critique an argument, you might need more information than

Classification of Changes in Matter
Which Changes Are Examples of a Chemical Change, and Which Are Examples of a Physical Change?

what is included on the whiteboard. You might, therefore, need to ask the presenter one or more follow-up questions, such as:

- What did your group do to analyze the data, and why did you decide to do it that way?
- Is that the only way to interpret the results of your group's analysis? How do you know that your interpretation of the analysis is appropriate?
- Why did your group decide to present your evidence in that manner?
- What other claims did your group discuss before deciding on that one? Why did you abandon those alternative ideas?
- How confident are you that your group's claim is valid? What could you do to increase your confidence?

Once the argumentation session is complete, you will have a chance to meet with your group and revise your original argument. Your group might need to gather more data or design a way to test one or more alternative claims as part of this process. Remember, your goal at this stage of the investigation is to develop the most valid or acceptable answer to the research question!

Report

Once you have completed your research, you will need to prepare an *investigation report* that consists of three sections that provide answers to the following questions:

1. What question were you trying to answer and why?
2. What did you do during your investigation and why did you conduct your investigation in this way?
3. What is your argument?

Your report should answer these questions in two pages or less. The report must be typed and any diagrams, figures, or tables should be embedded into the document. Be sure to write in a persuasive style; you are trying to convince others that your claim is acceptable or valid!

LAB 23

Checkout Questions

Lab 23. Classification of Changes in Matter: Which Changes Are Examples of a Chemical Change, and Which Are Examples of a Physical Change?

1. What are the characteristics of a physical change? What are the characteristics of a chemical change?

2. One night for dinner Jaxon decided to make baked potatoes for himself and his sister, Jade. He placed two potatoes in the oven set at 350°F for about 45 minutes. When Jaxon removed the potatoes from the oven, he noticed that they were soft and had a different texture than before they were cooked. Jade pointed out that the potatoes had undergone a chemical change. Jaxon did not believe that the potatoes had changed chemically; he thought they only experienced a physical change.

    Do you agree with Jade or with Jaxon? Use what you know about chemical and physical changes to provide a supporting argument.

Classification of Changes in Matter
Which Changes Are Examples of a Chemical Change, and Which Are Examples of a Physical Change?

3. Data and evidence are interchangeable in science.

   a. I agree with this statement.
   b. I disagree with this statement.

   Explain your answer, using an example from your investigation about classification of changes in matter.

4. In science, observations are objective, but inferences are subjective.

   a. I agree with this statement.
   b. I disagree with this statement.

   Explain your answer, using an example from your investigation about classification of changes in matter.

LAB 23

5. All things in the universe are made of matter. Understanding how matter moves within and between systems is important within science. Explain why understanding this is important, using an example from your investigation about classification of changes in matter.

6. When scientists observe events, often they are trying to recognize and identify patterns. Describe why patterns are important in science, using an example from your investigation about classification of changes in matter.

Lab Handout

Lab 24. Identification of Reaction Products: What Are the Products of the Chemical Reactions?

Introduction

Chemical reactions are the result of a rearrangement of the molecular or ionic structure of a substance. It is important to remember that the *law of conservation of mass* states that mass is conserved in ordinary chemical changes. The total amount of mass before and after the reaction is therefore the same, even though there are new substances with different properties than the original substances. Additionally, the *law of definite proportions* states that atoms combine in specific ways when they form compounds; therefore, a given compound always contains the same proportion of elements by mass. These two laws allow us to predict the rearrangement of atoms during chemical reactions, with no atoms being destroyed and no new atoms being produced. Balanced chemical equations are used to show the relative amounts of substances that react with each other and how the structures are rearranged during a chemical reaction.

FIGURE L24.1

An example of a precipitate reaction

One specific type of chemical reaction is a double replacement reaction or a *precipitation reaction*. Precipitation reactions typically occur when two solutions are mixed together and a nonsoluble product—the precipitate—is formed. Figure L24.1 shows an example of a precipitation reaction involving potassium iodide and lead nitrate. The balanced chemical equation is

$$2KI(aq) + Pb(NO_3)_2(aq) \rightarrow 2KNO_3(aq) + PbI_2(s)$$

In this example clear potassium iodide (KI) and clear lead nitrate ($Pb(NO_3)_2$) solutions are mixed together, producing a bright yellow precipitate, lead iodide (PbI_2). The other product, potassium nitrate (KNO_3), is soluble and remains dissolved in the solution.

To predict the products during a precipitation reaction, you must know the ion charges for the substances dissolved into the solutions and understand which types of substances are soluble in water. There are some general rules that can help you determine if an ionic compound will dissolve in water. Table L24.1 (p. 206) lists some basic solubility rules for ionic compounds; the table shows common *anions* (negatively charged ions) and *cations* (positively charged ions) along with their solubility. General solubility rules do not hold true in every case; therefore, exceptions to the solubility rules are also noted in Table L24.1.

LAB 24

TABLE L24.1
Solubility rules for ionic compounds in water

Ion	Soluble?	Exceptions
NO_3^-	Yes	None
ClO_4^-	Yes	None
Cl^-	Yes	Ag^+, Hg_2^{2+}, Pb^{2+}
I^-	Yes	Ag^+, Hg_2^{2+}, Pb^{2+}
SO_4^{2-}	Yes	Ca^{2+}, Ba^{2+}, Sr^{2+}, Ag^+, Hg^{2+}, Pb^{2+}
CO_3^{2-}	No	Group IA and NH_4^+
PO_4^{3-}	No	Group IA and NH_4^+
OH^-	No	Group IA, Ca^{2+} (slightly soluble), Ba^{2+}, Sr^{2+}
S^{2-}	No	Groups IA and IIA and NH_4^+
Na^+	Yes	None
NH_4^+	Yes	None
K^+	Yes	None

Understanding how ions may rearrange during a chemical reaction and understanding the general solubility rules will go a long way in helping you predict the products of a precipitation reaction and identify the actual precipitate. However, depending on the chemicals involved, there may be no obvious way to identify the precipitate using qualitative observations. In those cases it may be necessary to use stoichiometric procedures to determine the precipitate based on a balanced chemical equation.

Your Task

Four partial chemical equations are provided below. Your task is to identify the products of the four chemical reactions, including the precipitate in each reaction.

$$AgNO_3(aq) + NaCl(aq) \rightarrow ?$$

$$Na_2CrO_4(aq) + Ca(NO_3)_2(aq) \rightarrow ?$$

$$CaCl_2(aq) + Na_3PO_4(aq) \rightarrow ?$$

$$NaOH(aq) + NiCl_2(aq) \rightarrow ?$$

Identification of Reaction Products
What Are the Products of the Chemical Reactions?

The guiding question of this investigation is, **What are the products of the chemical reactions?**

Materials

You may use any of the following materials during your investigation:

Consumables	Equipment
• Calcium chloride, $CaCl_2$ • 1 M calcium nitrate, $Ca(NO_3)_2$ • 1 M nickel(II) chloride, $NiCl_2$ • 1 M silver nitrate, $AgNO_3$ • 1 M sodium chloride, $NaCl$ • 1 M sodium chromate, Na_2CrO_4 • 1 M sodium hydroxide, $NaOH$ • 1 M sodium phosphate, Na_3PO_4 • Distilled water	• Toothpicks • Filter paper • 4 test tubes • Well plate • Electronic or triple beam balance • Graduated cylinder (10 ml) • Vacuum filtration kit

Safety Precautions

Follow all normal lab safety rules. Silver nitrate is toxic by ingestion, is corrosive to body tissues, and stains clothes and skin. Your teacher will explain relevant and important information about working with the chemicals associated with this investigation. In addition, take the following safety precautions:

- Wear indirectly vented chemical-splash goggles and chemical-resistant gloves and apron while in the laboratory.
- Handle all glassware with care.
- Wash your hands with soap and water before leaving the laboratory.

Investigation Proposal Required?   ☐ Yes   ☐ No

Getting Started

To answer the guiding question, you will need to determine what type of data you need to collect, how you will collect the data, and how you will analyze the data.

To determine *what type of data you need to collect*, think about the following questions:

- How much of each chemical will you need to use?
- What masses will you need to measure during the investigation?
- What observations will you need to make?

To determine *how you will collect the data*, think about the following questions:

- How long will you need to allow the chemicals to react?

LAB 24

- How will you reduce error?

To determine *how you will analyze the data*, think about the following questions:

- What type of calculations will you need to make (if any)?
- How will you determine the precipitate in each reaction?

Connections to Crosscutting Concepts, the Nature of Science, and the Nature of Scientific Inquiry

As you work through your investigation, be sure to think about

- the importance of patterns in science,
- the importance of the flow of matter and energy within systems,
- the difference between observations and inferences in science, and
- the difference between laws and theories in science.

Initial Argument

Once your group has finished collecting and analyzing your data, you will need to develop an initial argument. Your argument must include a *claim*, which is your answer to the guiding question. Your argument must also include *evidence* in support of your claim. The evidence is your analysis of the data and your interpretation of what the analysis means. Finally, you must include a *justification* of the evidence in your argument. You will therefore need to use a scientific concept or principle to explain why the evidence that you decided to use is relevant and important. You will create your initial argument on a whiteboard. Your whiteboard must include all the information shown in Figure L24.2.

FIGURE L24.2
Argument presentation on a whiteboard

The Guiding Question:	
Our Claim:	
Our Evidence:	Our Justification of the Evidence:

Argumentation Session

The argumentation session allows all of the groups to share their arguments. One member of each group stays at the lab station to share that group's argument, while the other members of the group go to the other lab stations one at a time to listen to and critique the arguments developed by their classmates. The goal of the argumentation session is not to convince others that your argument is the best one; rather, the goal is to identify errors or instances of faulty reasoning in the initial arguments so these mistakes can be fixed. You will therefore need to evaluate the content of the claim, the quality of the evidence used to support the claim, and the strength of the justification of the evidence included in each argument that you see. To critique an argument, you might need more information than

Identification of Reaction Products
What Are the Products of the Chemical Reactions?

what is included on the whiteboard. You might, therefore, need to ask the presenter one or more follow-up questions, such as:

- What did your group do to analyze the data, and why did you decide to do it that way?
- Is that the only way to interpret the results of your group's analysis? How do you know that your interpretation of the analysis is appropriate?
- Why did your group decide to present your evidence in that manner?
- What other claims did your group discuss before deciding on that one? Why did you abandon those alternative ideas?
- How confident are you that your group's claim is valid? What could you do to increase your confidence?

Once the argumentation session is complete, you will have a chance to meet with your group and revise your original argument. Your group might need to gather more data or design a way to test one or more alternative claims as part of this process. Remember, your goal at this stage of the investigation is to develop the most valid or acceptable answer to the research question!

Report

Once you have completed your research, you will need to prepare an *investigation report* that consists of three sections that provide answers to the following questions:

1. What question were you trying to answer and why?
2. What did you do during your investigation and why did you conduct your investigation in this way?
3. What is your argument?

Your report should answer these questions in two pages or less. The report must be typed and any diagrams, figures, or tables should be embedded into the document. Be sure to write in a persuasive style; you are trying to convince others that your claim is acceptable or valid!

Checkout Questions

Lab 24. Identification of Reaction Products: What Are the Products of the Chemical Reactions?

1. Describe why a precipitate forms during some double displacement reactions and not others.

2. Stephanie and Tamara are conducting an investigation in chemistry class. They mix a solution of barium nitrate with a solution of sodium sulfate. When they mix the solutions, a reaction occurs and a precipitate is formed. Stephanie thinks the precipitate is sodium nitrate, but Tamara thinks it is barium sulfate.

    Use what you know about chemical reactions and solubility to provide an argument in support of either Stephanie or Tamara.

3. In science, observations are more important than inferences.

    a. I agree with this statement.
    b. I disagree with this statement.

    Explain your answer, using an example from your investigation about identification of reaction products.

Identification of Reaction Products
What Are the Products of the Chemical Reactions?

4. In science, there is a hierarchy of ideas that builds from hypothesis to theory to law, with each idea becoming more certain than the other.

    a. I agree with this statement.
    b. I disagree with this statement.

    Explain your answer, using an example from your investigation about identification of reaction products.

5. Scientists find it important to identify patterns in their observations. Explain why this is important, using an example from your investigation about identification of reaction products.

6. One of the main foci of chemistry is investigating the flow of matter and energy within systems. Explain why understanding the flow of matter and energy is important, using an example from your investigation about identification of reaction products.

LAB 25

Lab Handout

Lab 25. Acid-Base Titration and Neutralization Reactions: What Is the Concentration of Acetic Acid in Each Sample of Vinegar?

Introduction

Vinegar is basically a solution of acetic acid (CH_3COOH). It is commonly used as an ingredient in salad dressing and marinades. People also use it as a cleaning agent because it dissolves mineral deposits that often build up on appliances. The acetic acid found in vinegar is produced through the oxidation of ethanol (CH_3CH_2OH) by bacteria from the genus *Acetobacter* that is added to alcoholic liquids such as wine, apple cider, and beer. The overall chemical reaction that is facilitated by these bacteria is

$$CH_3CH_2OH + O_2 \rightarrow CH_3COOH + H_2O$$

The acetic acid concentration of vinegar differs depending on its intended use. The acetic acid concentration in table vinegars, such as red wine, apple cider or balsamic, ranges from 4% to 8%. When used for pickling, the acetic acid concentration of vinegar can be as high as 12%. Companies that produce vinegar therefore need to be able to determine the exact concentration of acetic acid in a sample of vinegar in order to ensure consistency between batches and to maintain quality control over their product. One way to determine the exact concentration of acid in a solution is to use a technique called an acid-base titration.

An acid-base titration is based on the premise that acids and bases neutralize each other when mixed in an exact stoichiometric ratio. For example, when sodium hydroxide is added to acetic acid, the sodium hydroxide will react with and consume the acetic acid based on the following neutralization reaction:

$$CH_3COOH(aq) + NaOH(aq) \rightarrow NaCH_3COO(aq) + H_2O(l)$$

This balanced chemical equation indicates that 1 mole of NaOH is needed to completely neutralize 1 mole of CH_3COOH. A chemist, as a result, can use this proportional relationship to determine how many moles of acetic acid are present in a solution. To accomplish this task, the chemist must first determine how many moles of sodium hydroxide need to be added to the solution in order to neutralize the acetic acid in it. The chemist can then use the stoichiometric ratio that exists between the reactants to determine the number of moles of acetic acid in the solution. Finally, the chemist can calculate the concentration of acetic acid in the solution based on the moles of acetic acid found in the sample and the volume of the sample.

Acid-Base Titration and Neutralization Reactions
What Is the Concentration of Acetic Acid in Each Sample of Vinegar?

In this investigation you will use an acid-base titration to determine the concentration of acetic acid in several different types of vinegar. This is important for you to be able to do because a common question that chemists have to answer is how much acid or base is present in a given solution. It is also a key aspect of doing acid-base chemistry.

Your Task
Determine the concentration of acetic acid in three different samples of vinegar.

The guiding question for this investigation is, **What is the concentration of acetic acid in each sample of vinegar?**

Materials
You may use any of the following materials during this investigation:

Consumables	Equipment
• Potassium hydrogen phthalate, KHP	• Volumetric pipette (5 ml)
• 0.1 M NaOH solution	• Pipette bulb
• Vinegar sample A	• Burette (50 ml)
• Vinegar sample B	• Burette clamp
• Vinegar sample C	• Test tube clamp
• Distilled water (in a wash bottle)	• Ring stand
Indicators	• 2 beakers (each 250 ml)
• Bromthymol blue	• 2 Erlenmeyer flasks (each 250 ml)
• Cresol red	• Graduated cylinder (25 ml)
• Phenolphthalein	• Magnetic stirrer or glass stirring rod
• Thymol blue	• Funnel
	• Electronic or triple beam balance

Safety Precautions
Follow all normal lab safety rules. Acetic acid and sodium hydroxide are corrosive to eyes, skin, and other body tissues. Sodium hydroxide is also toxic by ingestion. Phenolphthalein is an alcohol-based solution and is flammable, and it is moderately toxic by ingestion. Keep away from flames. Your teacher will explain relevant and important information about working with the chemicals associated with this investigation. In addition, take the following safety precautions:

- Wear indirectly vented chemical-splash goggles and chemical-resistant gloves and apron while in the laboratory.
- Handle all glassware with care.
- Wash your hands with soap and water before leaving the laboratory.

LAB 25

Investigation Proposal Required? ☐ Yes   ☐ No

Getting Started

The objective of an acid-base titration is to measure the volume of one reactant of known concentration that is required to neutralize another reactant of unknown concentration. The reactant with the known concentration is called the *titrant*, and the reactant with the unknown concentration is called the *analyte*. The titrant is gradually added to the analyte in small amounts until an end point is reached. The end point is represented by a color change in an indicator that is added to the analyte. If the indicator is chosen well, the end point will correspond with the equivalence point of reaction. The equivalence point is the point at which the amount of titrant is stoichiometrically equivalent to the amount of the analyte.

To perform an acid-base titration, you first need to understand how to measure the volume of the titrant needed to neutralize a specific volume of the analyte. The equipment setup for the process is illustrated in Figure L25.1. The basic steps are as follows:

1. Use a 5 ml volumetric pipette to add exactly 5 ml of the analyte (solution of unknown concentration) to a 250 ml beaker or 250 ml Erlenmeyer flask.
2. Add about 20 ml of deionized water to the analyte.
3. Add a few drops of indicator solution to the analyte.
4. Fill the burette with the titrant and record the initial burette reading.
5. Add the titrant drop by drop until the end point is reached. Be sure to stir or swirl the analyte as the titrant is added to make sure the chemicals are completely mixed.
6. Record the final burette reading.

Once the end point is reached, the volume of the titrant that was added to the analyte is used to calculate the molarity of the sample, as follows:

1. Determine the moles of titrant used in the reaction from the known molarity of the titrant and the volume of titrant needed to reach the end point.
2. Determine the moles of the analyte reactant using the stoichiometric ratio of the reactants provided in the neutralization equation and the number of moles of titrant used.
3. Determine the molarity of the unknown solution from the moles of the analyte reactant and the volume of the unknown sample (molarity = moles/L).

The concentration of the analyte can also be expressed as a mass percent. To determine the mass percent of the reactant in the analyte, perform the following calculations:

Acid-Base Titration and Neutralization Reactions
What Is the Concentration of Acetic Acid in Each Sample of Vinegar?

1. Use the moles of the analyte reactant and the molar mass of the analyte reactant to determine the mass of the reactant in the sample.

2. Determine the total mass of the sample.

3. Determine the mass percent of the analyte reactant from the mass of the analyte reactant in the sample and the total mass of the sample.

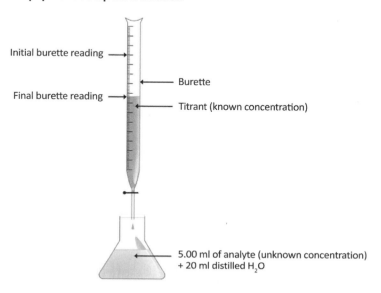

FIGURE L25.1

Equipment setup for a titration

When performing a titration, it is important to know the exact concentration of the titrant. Solid NaOH, for example, tends to absorb moisture from the air. It is therefore often difficult to determine the exact mass of NaOH that is added to solution, which leads to inaccurate molarity calculations. In addition, solutions of NaOH tend to absorb carbon dioxide from the air, which neutralizes some of the base. NaOH should therefore be standardized before it is used as the titrant in a titration. You instructor will either tell you the standardized concentration of the NaOH solution that you will use during this investigation or have you standardize your NaOH solution using a primary standard called potassium hydrogen phthalate (KHP).

It is also important to choose an indicator solution that will signal the end point of the titration that is as close as possible to the equivalence point of the reaction. Acetic acid is a weak acid. The equivalence point of the reaction between acetic acid and sodium hydroxide is at a pH of 8.72. KHP is also a weak acid. The equivalence point of the reaction between KHP and sodium hydroxide is at a pH of 8.50. Table L25.1 provides a list of the potential indicators that you can use for this investigation.

TABLE L25.1
Potential indicators

Indicator	pH range	Low pH color	High pH color
Bromthymol blue	6.0–7.6	Yellow	Blue
Cresol red	7.2–8.8	Orange	Red
Phenolphthalein	8.0–9.6	Clear	Pink
Thymol blue	8.0–9.6	Yellow	Blue

LAB 25

Now that you understand how to perform a titration, you will need to determine what type of data you need to collect, how you will collect the data, and how you will analyze the data to answer the guiding question.

To determine *what type of data you need to collect*, think about the following questions:

- What type of measurements will you need to record during your investigation?
- When will you need to take your measurements?

To determine *how you will collect the data*, think about the following questions:

- What samples of vinegar will you need to titrate?
- Which indicator will you use and why?
- How will you eliminate confounding variables?
- How will you reduce measurement error? (Hint: You can run multiple trials and then determine if the results are significantly close. You will also need to average your results.)
- How will you keep track of the data you collect?

To determine *how you will analyze the data*, think about the following questions:

- What types of calculations will you need to make to determine the concentration of acetic acid in each sample of vinegar?
- How will you determine which sample of vinegar has the greatest concentration?
- What type of graph could you create to help make sense of your data?

Connections to Crosscutting Concepts, the Nature of Science, and the Nature of Scientific Inquiry

As you work through your investigation, be sure to think about

- the importance of recognizing and using proportional relationships,
- the importance of tracking how matter moves into and within a system,
- the difference between observations and inferences in science, and
- how science is influenced by the society and culture in which it is practiced.

Initial Argument

Once your group has finished collecting and analyzing your data, you will need to develop an initial argument. Your argument must include a *claim*, which is your answer to the guiding question. Your argument must also include *evidence* in support of your claim. The evidence is your analysis of the data and your interpretation of what the analysis means. Finally, you must include a *justification* of the evidence in your argu-

Acid-Base Titration and Neutralization Reactions
What Is the Concentration of Acetic Acid in Each Sample of Vinegar?

ment. You will therefore need to use a scientific concept or principle to explain why the evidence that you decided to use is relevant and important. You will create your initial argument on a whiteboard. Your whiteboard must include all the information shown in Figure L25.2.

Argumentation Session

The argumentation session allows all of the groups to share their arguments. One member of each group stays at the lab station to share that group's argument, while the other members of the group go to the other lab stations one at a time to listen to and critique the arguments developed by their classmates. The goal of the argumentation session is not to convince others that your argument is the best one; rather, the goal is to identify errors or instances of faulty reasoning in the initial arguments so these mistakes can be fixed. You will therefore need to evaluate the content of the claim, the quality of the evidence used to support the claim, and the strength of the justification of the evidence included in each argument that you see. To critique an argument, you might need more information than what is included on the whiteboard. You might, therefore, need to ask the presenter one or more follow-up questions, such as:

FIGURE L25.2
Argument presentation on a whiteboard

The Guiding Question:	
Our Claim:	
Our Evidence:	Our Justification of the Evidence:

- What did your group do to make sure the data you collected are reliable? What did you do to decrease measurement error?
- What did your group do to analyze the data, and why did you decide to do it that way? Did you check your calculations?
- Is that the only way to interpret the results of your group's analysis? How do you know that your interpretation of the analysis is appropriate?
- Why did your group decide to present your evidence in that manner?
- What other claims did your group discuss before deciding on that one? Why did you abandon those alternative ideas?
- How confident are you that your group's claim is valid? What could you do to increase your confidence?

Once the argumentation session is complete, you will have a chance to meet with your group and revise your original argument. Your group might need to gather more data or design a way to test one or more alternative claims as part of this process. Remember, your goal at this stage of the investigation is to develop the most valid or acceptable answer to the research question!

LAB 25

Report

Once you have completed your research, you will need to prepare an *investigation report* that consists of three sections that provide answers to the following questions:

1. What question were you trying to answer and why?
2. What did you do during your investigation and why did you conduct your investigation in this way?
3. What is your argument?

Your report should answer these questions in two pages or less. The report must be typed and any diagrams, figures, or tables should be embedded into the document. Be sure to write in a persuasive style; you are trying to convince others that your claim is acceptable or valid!

Acid-Base Titration and Neutralization Reactions
What Is the Concentration of Acetic Acid in Each Sample of Vinegar?

Reference Sheet

Standardization of a NaOH Solution

A solution of potassium hydrogen phthalate ($KHC_8H_4O_4$), often called KHP, can be used to standardize a NaOH solution. KHP serves as a primary standard, which is a stable solid that can be weighed out and used to standardize a titrant solution, such as NaOH, that has a concentration that may be slightly different from the reported value.

The balanced chemical equation for the neutralization of KHP by NaOH is

$$KHC_8H_4O_4(aq) + NaOH(aq) \rightarrow KNaC_8H_4O_4(aq) + H_2O(l)$$

This equation indicates that the stoichiometric ratio of NaOH and KHP is 1:1. Therefore if the number of grams of solid KHP used to make the solution of KHP is known, it is easy to determine the number of moles of NaOH used in the titration and the exact concentration of the NaOH solution. The basic steps are as follows:

1. Weigh out about 0.5 g of KHP and record the exact mass.

2. Transfer the KHP to a 250 ml Erlenmeyer flask and add exactly 50 ml of distilled water.

3. Add a few drops of indicator solution to the KHP solution.

4. Fill the burette with the NaOH solution and record the initial burette reading.

5. Add the titrant drop by drop until the end point is reached. Be sure to stir or swirl the KHP solution as the titrant is added to make sure the chemicals are completely mixed.

6. Record the final burette reading.

Once the end point is reached, the volume of the NaOH that was added to the KHP solution can be used to calculate the molarity of the NaOH, as follows:

1. Determine the moles of KHP used (molar mass of KHP is 204.23 g/mol).

2. Determine the moles of NaOH used (moles of NaOH = moles of KHP).

3. Determine the molarity of the NaOH solution (molarity = moles of NaOH used in the titration / liters of NaOH used in the titration).

Be sure to complete at least three trials to determine the amount of measurement error associated with the titration. If the results are sufficiently close to each other, use the average value for the calculations.

Checkout Questions

Lab 25. Acid-Base Titration and Neutralization Reactions: What Is the Concentration of Acetic Acid in Each Sample of Vinegar?

1. Why can chemists use an acid-base titration to determine the concentration of an unknown acid or base?

2. If you titrate 20.0 ml of 5 M sodium hydroxide with 5.0 ml of hydrochloric acid, what is the molarity of the acid? Hydrochloric acid reacts with sodium hydroxide as follows: $HCl + NaOH \rightarrow NaCl + H_2O$.

3. All scientists will make the same observations during an investigation.

    a. I agree with this statement.
    b. I disagree with this statement.

   Explain your answer, using an example from your investigation about acid-base titration and neutralization reactions.

Acid-Base Titration and Neutralization Reactions
What Is the Concentration of Acetic Acid in Each Sample of Vinegar?

4. The research done by a scientist is often influenced by what is important in society.

    a. I agree with this statement.
    b. I disagree with this statement.

    Explain your answer, using an example from your investigation about acid-base titration and neutralization reactions.

5. Scientists often need to track how matter moves into and within a system. Explain why this is important, using an example from your investigation about the concentration of acetic acid in vinegar.

6. Scientists often focus on proportional relationships. Explain what a proportional relationship is and why these relationships are useful, using an example from your investigation about acid-base titration and neutralization reactions.

Lab Handout

Lab 26. Composition of Chemical Compounds: What Is the Empirical Formula of Magnesium Oxide?

Introduction

Chemists can describe the composition of a chemical compound in at least three different ways. The first way is to define the *percent composition* of a compound; this is the percent of each element found in the compound by mass. For example, a chemical compound called acetylene is composed of 92.25% carbon and 7.75% hydrogen by mass. The second way to describe the composition of a compound is to provide its *empirical formula*, which indicates the lowest whole-number ratio of the atoms found within one unit of that compound. The empirical formula of acetylene, for example, is CH. The empirical formula indicates the ratio of atoms in a compound but does not always represent the actual number of each kind of atom found in one unit of that compound. The compound benzene, for example, has the same empirical formula (CH) as acetylene because both acetylene and benzene contain one carbon atom for every atom of hydrogen. The third way to define the percent composition of a compound is to use a *molecular formula*, which indicates the actual number of atoms that are found in a single unit of that compound. The molecular formulas of acetylene and benzene are different even though they share the same empirical formula because each compound contains a different number of carbon and hydrogen atoms. The molecular formula of acetylene is C_2H_2, whereas the molecular formula for benzene is C_6H_6.

Chemists rely on two important principles when they attempt to determine the composition of an unknown compound. The first principle is the *law of definite proportions*, which indicates that a compound is always made up of the exact same proportion of elements by mass. The percent composition of a compound is therefore a constant and does not depend on the amount of a sample. The second principle is the *law of conservation of mass*, which states that mass is neither created nor destroyed during a chemical reaction. These two principles enable chemists to determine the percent composition of an unknown compound. Chemists can then use the percent composition of the compound and some simple mathematics to determine its empirical formula. In this investigation, you will have an opportunity to use these two principles to determine the empirical formula of a compound that you create inside the lab by heating magnesium in the presence of oxygen.

Your Task

Magnesium oxide is a compound that consists of magnesium and oxygen. It is produced when magnesium metal is heated. The heat causes the magnesium to combine with molecules of oxygen found in the air. The magnesium and oxygen may combine in a number

Composition of Chemical Compounds
What Is the Empirical Formula of Magnesium Oxide?

of different ratios during this reaction. These ratios include, but are not limited to, MgO, Mg_2O, Mg_3O_2, and Mg_5O_4. Your goal is to determine the percent composition of magnesium oxide and then use this information to calculate its empirical formula.

The guiding question for this lab is, **What is the empirical formula of magnesium oxide?**

Materials

You may use any of the following materials during your investigation:

Consumables	Equipment
• Magnesium ribbon (3–4 cm long)	• Bunsen burner
	• Striker
	• Crucible with lid
	• Clay triangle
	• Crucible tongs
	• Ring stand with metal ring
	• Wire gauze square
	• Electronic or triple beam balance
	• Periodic table

Safety Precautions

Follow all normal lab safety rules. Your teacher will explain relevant and important information about working with the chemicals associated with this investigation. In addition, take the following safety precautions:

- Wear indirectly vented chemical-splash goggles and chemical-resistant gloves and apron while in the laboratory.
- Use caution when working with Bunsen burners. They can burn skin, and combustibles and flammables must be kept away from the open flame. If you have long hair, tie it back behind your head.
- Inspect the crucible for cracks. If it is cracked, exchange it for a new one. Clean the crucible and lid thoroughly before using them.
- Be careful with a crucible after removing it from a flame because it will still be hot.
- Handle all glassware with care.
- Wash your hands with soap and water before leaving the laboratory.

Investigation Proposal Required?   ☐ Yes   ☐ No

Getting Started

The first step in your investigation is to determine the percent composition of magnesium oxide. As noted earlier, when magnesium is heated it, combines with oxygen to form magnesium oxide. The law of conservation of mass suggests that the total mass of the products of a chemical reaction must equal the mass of the reactants. The mass of the oxygen in a sample of magnesium oxide will therefore equal the mass of the magnesium oxide minus the mass of the original piece of magnesium that was used in the reaction. You can use this information to determine the percent of oxygen by mass found in magnesium oxide. You will, however, need to first determine what type of data you need to collect and how you will collect the data to be able to make this calculation.

To determine *what type of data you need to collect*, think about the following questions:

- What type of measurements or observations will you need to record during your investigation?
- When will you need to make these measurements or observations?

To determine *how you will collect the data*, think about the following questions:

- What equipment will you need to use to make magnesium oxide?
- How much magnesium will you use to make magnesium oxide?
- How will you know when all the magnesium has been converted to magnesium oxide?
- How will you make sure that your data are of high quality (i.e., how will you reduce error)?
- How will you keep track of the data you collect and how will you organize it?

The second step in your investigation is to calculate the empirical formula for magnesium oxide from its percent composition. This calculation requires two steps. First, you will need to calculate the number of moles of magnesium and oxygen in your sample of magnesium oxide. Next, you will need to use the ratio between the number of moles of magnesium and the number of moles of oxygen to calculate the empirical formula of magnesium oxide. Keep in mind that fractions of atoms do not exist.

Connections to Crosscutting Concepts, the Nature of Science, and the Nature of Scientific Inquiry

As you work through your investigation, be sure to think about

- how scale, proportion, and quantity play a role in science;
- the flow of energy and matter within systems;
- how scientific knowledge can change over time in light of new evidence; and
- the different methods used in scientific investigations.

Composition of Chemical Compounds
What Is the Empirical Formula of Magnesium Oxide?

Initial Argument

Once your group has finished collecting and analyzing your data, you will need to develop an initial argument. Your argument must include a *claim*, which is your answer to the guiding question. Your argument must also include *evidence* in support of your claim. The evidence is your analysis of the data and your interpretation of what the analysis means. Finally, you must include a *justification* of the evidence in your argument. You will therefore need to use a scientific concept or principle to explain why the evidence that you decided to use is relevant and important. You will create your initial argument on a whiteboard. Your whiteboard must include all the information shown in Figure L26.1.

FIGURE L26.1
Argument presentation on a whiteboard

The Guiding Question:	
Our Claim:	
Our Evidence:	Our Justification of the Evidence:

Argumentation Session

The argumentation session allows all of the groups to share their arguments. One member of each group stays at the lab station to share that group's argument, while the other members of the group go to the other lab stations one at a time to listen to and critique the arguments developed by their classmates. The goal of the argumentation session is not to convince others that your argument is the best one; rather, the goal is to identify errors or instances of faulty reasoning in the initial arguments so these mistakes can be fixed. You will therefore need to evaluate the content of the claim, the quality of the evidence used to support the claim, and the strength of the justification of the evidence included in each argument that you see. To critique an argument, you might need more information than what is included on the whiteboard. You might, therefore, need to ask the presenter one or more follow-up questions, such as:

- How did your group collect the data? Why did you use that method?
- Is that the only way to interpret the results of your group's analysis? How do you know that your interpretation of the analysis is appropriate?
- Why did your group decide to present your evidence in that manner?
- What other claims did your group discuss before deciding on that one? Why did you abandon those alternative ideas?
- How confident are you that your group's claim is valid? What could you do to increase your confidence?

Once the argumentation session is complete, you will have a chance to meet with your group and revise your original argument. Your group might need to gather more data or design a way to test one or more alternative claims as part of this process. Remember, your goal at this stage of the investigation is to develop the most valid or acceptable answer to the research question!

Report

Once you have completed your research, you will need to prepare an *investigation report* that consists of three sections that provide answers to the following questions:

1. What question were you trying to answer and why?
2. What did you do during your investigation and why did you conduct your investigation in this way?
3. What is your argument?

Your report should answer these questions in two pages or less. The report must be typed and any diagrams, figures, or tables should be embedded into the document. Be sure to write in a persuasive style; you are trying to convince others that your claim is acceptable or valid!

Composition of Chemical Compounds
What Is the Empirical Formula of Magnesium Oxide?

Checkout Questions

Lab 26. Composition of Chemical Compounds: What Is the Empirical Formula of Magnesium Oxide?

1. Describe how the law of definite proportions and the law of conservation of mass are useful for determining the composition of a compound.

2. Felicity and Dawson are conducting an investigation to determine the empirical formula of iron oxide. They start with an 85.65 g piece of iron metal and burn it in air. The mass of the iron oxide produced is 118.37. Felicity thinks the empirical formula of iron oxide is Fe_3O_4 and Dawson thinks it is FeO. Use what you know about how to determine the empirical formula of a compound to provide an argument in support of Felicity or Dawson.

3. If multiple scientists are investigating the same question in their own labs, each scientist will use the same method to try and answer the question.

    a. I agree with this statement.
    b. I disagree with this statement.

    Explain your answer, using an example from your investigation about the composition of chemical compounds.

4. In science, once an idea has been established it will not be changed because it has been investigated so thoroughly.

    a. I agree with this statement.
    b. I disagree with this statement.

LAB 26

Explain your answer, using an example from your investigation about the composition of chemical compounds.

5. Scale, proportion, and quantity play a central role in science and chemistry. Explain why these concepts are important, using an example from your investigation about the composition of chemical compounds.

6. One of the main foci of chemistry is investigating the flow of matter and energy within systems. Explain how understanding the flow of matter and energy can be useful, using an example from your investigation about the composition of chemical compounds.

Lab Handout

Lab 27. Stoichiometry and Chemical Reactions: Which Balanced Chemical Equation Best Represents the Thermal Decomposition of Sodium Bicarbonate?

Introduction

The *law of conservation of mass* states that mass is conserved during a chemical reaction. The *law of definite proportions* states that a compound is always made up of the exact same proportion of elements by mass. John Dalton was able to explain these two fundamental laws of chemistry with his *atomic theory*, which states that a chemical reaction is simply the rearrangement of atoms with no atoms being destroyed and no new atoms being produced during the process. Chemists use a balanced chemical equation to represent what happens on the submicroscopic level during a chemical reaction.

The *stoichiometric coefficient* is the number written in front of atoms, ions, or molecules in a chemical equation. These numbers are used to balance the number of each type of atom found on both the reactant and product sides of the equation. Stoichiometric coefficients are also useful because they identify the mole ratio between reactants and products. The mole ratio is important because it allows chemists to determine how many moles of a product will be produced from a specific number of moles of a reactant or how many moles of reactant are needed to produce a specific amount of a product.

Molar mass serves as a bridge between the number of moles of a substance and the mass of a substance. The molar mass is the mass of a given substance divided by one mole of the substance. The molar mass of a given substance can be calculated by summing the atomic mass for each atom found in a molecule of that substance. For example, the atomic mass of hydrogen is 1.01 g/mol, and the atomic mass of oxygen is 15.99 g/mol, so the molar mass of H_2O is 18.01 g/mol (1.01 g/mol + 1.01 g/mol + 15.99 g/mol). Once the molar mass of a substance is known, the mass of a sample can be used to determine the number of moles of a substance or the moles of substance can be used to determine the mass of a sample. For example, a 40-gram sample of H_2O consists of 2.2 mol of H_2O (40 g of H_2O ÷ 18.01 g/mol = 2.2 mol of H_2O) and a 3.0 mol sample of H_2O has a mass of 54.03 g (3.0 mol of H_2O × 18.01 g/mol = 54.03 g of H_2O). In this investigation, you will have an opportunity to use atomic theory, molar mass, and stoichiometry to determine how atoms are rearranged during a chemical reaction.

LAB 27

Your Task

There are at least four different balanced chemical equations that could explain how atoms are rearranged during the thermal decomposition of sodium bicarbonate ($NaHCO_3$). The first potential explanation is that the sodium bicarbonate decomposes into sodium hydroxide (NaOH) and carbon dioxide (CO_2) when it is heated. The balanced chemical equation for this reaction is

$$NaHCO_3(s) \rightarrow NaOH(s) + CO_2(g)$$

The second potential explanation is that the sodium bicarbonate decomposes into sodium carbonate (Na_2CO_3), carbon dioxide (CO_2), and water when it is heated. The balanced chemical equation for this reaction is

$$2NaHCO_3(s) \rightarrow Na_2CO_3(s) + CO_2(g) + H_2O(g)$$

The third potential explanation is that the sodium bicarbonate decomposes into sodium oxide (Na_2O), carbon dioxide, and water when it is heated. The balanced chemical equation for this potential reaction is

$$2NaHCO_3(s) \rightarrow Na_2O(s) + 2CO_2(g) + H_2O(g)$$

The fourth potential explanation is that the sodium bicarbonate decomposes into sodium hydride (NaH), carbon monoxide (CO), and oxygen when it is heated. The balanced chemical equation for this potential reaction is

$$NaHCO_3(s) \rightarrow NaH(s) + CO(g) + O_2(g)$$

Your goal is to determine which of these four balanced chemical equations best represents how atoms are rearranged during the thermal decomposition of sodium bicarbonate.

The guiding question of this investigation is, **Which balanced chemical equation best represents the thermal decomposition of sodium bicarbonate?**

Materials

You may use any of the following materials during your investigation:

Consumable	Equipment
Solid $NaHCO_3$	• Bunsen burner • Striker • Ring stand with metal ring • Crucible with lid • Crucible tongs • Pipe-stem triangle • Wire gauze square • Electronic or triple beam balance • Periodic table

Stoichiometry and Chemical Reactions
Which Balanced Chemical Equation Best Represents the Thermal Decomposition of Sodium Bicarbonate?

Safety Precautions

Follow all normal lab safety rules. Your teacher will explain relevant and important information about working with the chemicals associated with this investigation. In addition, take the following safety precautions:

- Wear indirectly vented chemical-splash goggles and chemical-resistant gloves and apron while in the laboratory.
- Use caution when working with Bunsen burners. They can burn skin, and combustibles and flammables must be kept away from the open flame. If you have long hair, tie it back behind your head.
- Inspect the crucible for cracks. If it is cracked, exchange it for a new one. Clean the crucible and lid thoroughly before using them.
- Be careful with a crucible after removing it from a flame because it will still be hot.
- Wash your hands with soap and water before leaving the laboratory.

Investigation Proposal Required? ☐ Yes ☐ No

Getting Started

As part of your investigation, you will need to use a Bunsen burner and a crucible (see Figure L27.1) to increase the temperature of sodium bicarbonate enough for it to decompose. The thermal decomposition of sodium bicarbonate will occur rapidly at 200°C, but the product of the decomposition reaction will begin to decompose at temperatures over 850°C.

FIGURE L27.1

How to heat sodium bicarbonate using a crucible and a Bunsen burner

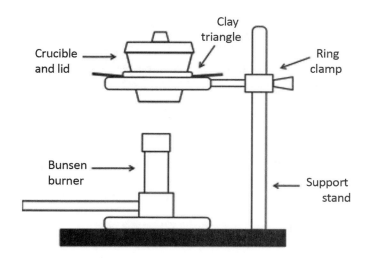

LAB 27

To answer the guiding question, you will also need to determine what type of data you will need to collect during your investigation, how you will collect the data, and how you will analyze the data.

To determine *what type of data to collect*, think about the following questions:

- How much $NaHCO_3$ will you need to use?
- What will you need to measure?

To determine *how you will collect the data*, think about the following questions:

- How long will you need to heat the $NaHCO_3$?
- How will you empirically determine when the decomposition of the $NaHCO_3$ is complete?
- How will you reduce error?

To determine *how you will analyze the data*, think about the following questions:

- What type of calculations will you need to make?
- How will your group take into account the precision of the balance in your analysis?

Connections to Crosscutting Concepts, the Nature of Science, and the Nature of Scientific Inquiry

As you work through your investigation, be sure to think about

- the importance of identifing the underlying cause for observed phenomena,
- how models are used to study natural phenomena,
- the difference between data and evidence in science, and
- the nature and role of experiments in science.

Initial Argument

Once your group has finished collecting and analyzing your data, you will need to develop an initial argument. Your argument must include a *claim*, which is your answer to the guiding question. Your argument must also include *evidence* in support of your claim. The evidence is your analysis of the data and your interpretation of what the analysis means. Finally, you must include a *justification* of the evidence in your argument. You will therefore need to use a scientific concept or principle to explain why the evidence that you decided to use is relevant and important. You will create your initial argument on a whiteboard. Your whiteboard must include all the information shown in Figure L27.2.

Stoichiometry and Chemical Reactions
Which Balanced Chemical Equation Best Represents the Thermal Decomposition of Sodium Bicarbonate?

Argumentation Session

The argumentation session allows all of the groups to share their arguments. One member of each group stays at the lab station to share that group's argument, while the other members of the group go to the other lab stations one at a time to listen to and critique the arguments developed by their classmates. The goal of the argumentation session is not to convince others that your argument is the best one; rather, the goal is to identify errors or instances of faulty reasoning in the initial arguments so these mistakes can be fixed. You will therefore need to evaluate the content of the claim, the quality of the evidence used to support the claim, and the strength of the justification of the evidence included in each argument that you see. To critique an argument, you might need more information than what is included on the whiteboard. You might, therefore, need to ask the presenter one or more follow-up questions, such as:

FIGURE L27.2
Argument presentation on a whiteboard

The Guiding Question:	
Our Claim:	
Our Evidence:	Our Justification of the Evidence:

- How did your group collect the data? Why did you use that method?
- What did your group do to make sure the data you collected are reliable? What did you do to decrease measurement error?
- What did your group do to analyze the data, and why did you decide to do it that way? Did you check your calculations?
- Is that the only way to interpret the results of your group's analysis? How do you know that your interpretation of the analysis is appropriate?
- Why did your group decide to present your evidence in that manner?
- What other claims did your group discuss before deciding on that one? Why did you abandon those alternative ideas?
- How confident are you that your group's claim is valid? What could you do to increase your confidence?

Once the argumentation session is complete, you will have a chance to meet with your group and revise your original argument. Your group might need to gather more data or design a way to test one or more alternative claims as part of this process. Remember, your goal at this stage of the investigation is to develop the most valid or acceptable answer to the research question!

Report

Once you have completed your research, you will need to prepare an *investigation report* that consists of three sections that provide answers to the following questions:

LAB 27

1. What question were you trying to answer and why?
2. What did you do during your investigation and why did you conduct your investigation in this way?
3. What is your argument?

Your report should answer these questions in two pages or less. The report must be typed and any diagrams, figures, or tables should be embedded into the document. Be sure to write in a persuasive style; you are trying to convince others that your claim is acceptable or valid!

Checkout Questions

Lab 27. Stoichiometry and Chemical Reactions: Which Balanced Chemical Equation Best Represents the Thermal Decomposition of Sodium Bicarbonate?

1. Describe how the law of definite proportions is useful in understanding chemical reactions and predicting their products.

2. Alex and Sam are conducting an investigation to determine the products of the decomposition of barium carbonate ($BaCO_3$). They started with 8.0 g of $BaCO_3$, which was heated and decomposed, leaving 6.5 g of solid product.

    Alex thinks the correct chemical reaction is
    $$BaCO_3(s) \rightarrow BaO(s) + CO_2(g)$$
    But Sam thinks the correct reaction is
    $$BaCO_3(s) \rightarrow BaCO(s) + O_2(g)$$

    Use what you know about chemical reactions to provide an argument in support of Alex or Sam, based on the data provided.

LAB 27

3. Experiments are the best way that scientists can learn about the natural world.

    a. I agree with this statement.
    b. I disagree with this statement.

    Explain your answer, using an example from your investigation about stoichiometry and chemical reactions.

4. In science, evidence is more important than data.

    a. I agree with this statement.
    b. I disagree with this statement.

    Explain your answer, using an example from your investigation about stoichiometry and chemical reactions.

5. Scientists make many observations, and they also propose causal mechanisms that may underlie their observations. Explain why understanding causal mechanisms is important, using an example from your investigation about stoichiometry and chemical reactions.

6. A chemical equation can be considered a model for a chemical reaction. Explain what a model is and why models are important in science, using an example from your investigation about stoichiometry and chemical reactions.

Lab Handout

Lab 28. Designing a Cold Pack: Which Salt Should Be Used to Make an Effective but Economical Cold Pack?

Introduction

An instant cold pack is a first aid device that is used to treat injuries. Most commercial instant cold packs contain two plastic bags. One bag contains an ionic compound, and the other bag contains water. When the instant cold pack is squeezed hard enough, the bag containing the water breaks and the ionic compound and water mix. The dissolution of the ionic compound in the water results in an enthalpy change and a decrease in the overall temperature of the cold pack. In this investigation, you will explore the enthalpy changes that are associated with common salts and then apply what you have learned about these enthalpy changes to design an effective but economical instant cold pack.

The enthalpy change associated with the dissolution process is called the *heat of solution* (ΔH_{soln}). At constant pressure, the ΔH_{soln} is equal in magnitude to heat (q) lost to or gained from the surroundings. In the case of a salt dissolving in water, the overall enthalpy change is the net result of two key processes. First, an input of energy is required to break the attractive forces that hold the ions in the salt together and to disrupt the intermolecular forces that hold the water molecules in the solvent together. The system *gains* energy during this process. Second, energy is released from the system as attractive forces form between the dissociated ions and the molecules of water. The system *loses* energy during this process. The ΔH_{soln} can therefore be either endothermic or exothermic depending on the net energy change in the system. The ΔH_{soln} is exothermic when the system releases more energy into the surroundings than it absorbs and endothermic when the system absorbs more energy than it releases.

A chemist can determine the molar ΔH_{soln} for a specific salt by mixing a sample of it with water inside a *calorimeter*. A calorimeter is an insulated container that is designed to prevent or at least reduce heat loss to the atmosphere (see Figure L28.1, p. 238). Once the salt and water are mixed, the chemist can record the temperature change that occurs inside the calorimeter as a result of the dissolution process. The magnitude of the heat energy change is then calculated using the following equation:

$$q = m \times s \times \Delta T$$

where q = heat energy change (in joules), m = total mass of the solution (solute plus solvent), s = the specific heat of the solution (4.18 J/g•°C), and ΔT = the observed temperature change. The chemist can then calculate the molar ΔH_{soln} for the salt by dividing q by the number of moles of the salt (n) that he or she mixed with the water.

LAB 28

Your Task

Investigate different salts for potential use in a cold pack. Using the empirical data you collect along with the cost data provided in Table L28.1 (p. 240), determine which salt in what quantity should be used to produce an effective but economical cold pack.

The guiding question of this investigation is, **Which salt should be used to make an effective but economical cold pack?**

Materials

You may use any of the following materials during your investigation:

Consumables	Equipment
• Ammonium chloride, NH_4Cl • Ammonium nitrate, NH_4NO_3 • Magnesium sulfate, $MgSO_4$ • Sodium thiosulfate, $Na_2S_2O_3$ • Distilled water	• Graduated cylinder (100 ml) • Spatula • Calorimeter • Temperature probe with sensor interface • Electronic or triple beam balance

FIGURE L28.1
Example of a calorimeter

Safety Precautions

Follow all normal lab safety rules. Ammonium chloride, ammonium nitrate, sodium thiosulfate, and magnesium sulfate are all tissue irritants and moderately toxic by ingestion. Your teacher will explain relevant and important information about working with the chemicals associated with this investigation. In addition, take the following safety precautions:

- Wear indirectly vented chemical-splash goggles and chemical-resistant gloves and apron while in the laboratory.
- Handle all glassware with care.
- Wash your hands with soap and water before leaving the laboratory.

Investigation Proposal Required?   ☐ Yes   ☐ No

Getting Started

The first step in your investigation is to determine the heat energy change associated with each salt. To accomplish this task, you will need to determine what type of data to collect, how you will collect the data, and how you will analyze the data.

To determine *what type of data you need to collect*, think about the following questions:

Designing a Cold Pack
Which Salt Should Be Used to Make an Effective but Economical Cold Pack?

- What type of measurements or observations will you need to make during your investigation?
- Is it important to know the change in temperature of the solution or just its final temperature?
- How does the amount of salt or the amount of water influence your potential results?

To determine *how you will collect the data*, think about the following questions:

- What will serve as your independent and dependent variables?
- How often will you collect data and when will you do it?
- How will you make sure that your data are of high quality (i.e., how will you reduce error)?
- How will you keep track of the data you collect and how will you organize it?

To determine *how you will analyze the data*, think about the following questions:

- How will you calculate the heat energy change associated with the formation of a solution?
- How will you calculate the molar ΔH_{soln} for each compound?
- What type of graph could you create to help make sense of your data?

The second step of your investigation will be to determine which salt should be used to make the instant cold pack. The company wants to produce small instant cold packs that will easily fit in a portable first aid kit. The instant cold pack they are planning to make will consist of two bags: one containing water and the other containing one of the salts. The bag of water will be placed inside the bag that contains the salt so when the bag of water is ruptured, the salt and water can mix. The company is planning on using 60 ml of water in this cold pack. For the instant ice pack to be effective, its temperature needs to fall to about 2°C once the salt and water are mixed. The company, however, wants to spend as little as possible to produce the instant cold packs. You will therefore need to conduct a complete cost-benefit analysis for each salt. This will require you to determine how much of each type of salt you will need to use and how much it will cost per instant cold pack. The price of each salt is given in Table L28.1 (p. 240).

LAB 28

TABLE L28.1
Prices of salts

Salt	Amount (in grams)	Price
NH_4Cl	1,000	$13.90
NH_4NO_3	500	$8.95
$MgSO_4$	100	$1.17
$Na_2S_2O_3$	500	$8.55

Connections to Crosscutting Concepts, the Nature of Science, and the Nature of Scientific Inquiry

As you work through your investigation, be sure to think about

- the importance of recognizing and analyzing patterns,
- how energy and matter flow within a system,
- the role of culture and values in science, and
- the importance of creativity in science.

Initial Argument

Once your group has finished collecting and analyzing your data, you will need to develop an initial argument. Your argument must include a *claim*, which is your answer to the guiding question. Your argument must also include *evidence* in support of your claim. The evidence is your analysis of the data and your interpretation of what the analysis means. Finally, you must include a *justification* of the evidence in your argument. You will therefore need to use a scientific concept or principle to explain why the evidence that you decided to use is relevant and important. You will create your initial argument on a whiteboard. Your whiteboard must include all the information shown in Figure L28.2.

FIGURE L28.2
Argument presentation on a whiteboard

The Guiding Question:	
Our Claim:	
Our Evidence:	Our Justification of the Evidence:

Argumentation Session

The argumentation session allows all of the groups to share their arguments. One member of each group stays at the lab station to share that group's argument, while the other members of the group go to the other lab stations one at a time to listen to and critique the arguments developed by their classmates. The goal of the argumentation session is not to convince others that your argument is the best one; rather, the goal is to identify errors or instances of faulty reasoning in the initial arguments so these mistakes can be fixed. You will therefore need to evaluate the content of the claim, the quality of the evidence used to support the claim, and the strength of the justification of the evidence included in each argument that you see. To critique an argument, you might need more information than what is included on the whiteboard. You might, therefore, need to ask the presenter one or more follow-up questions, such as:

- How did your group collect the data? Why did you use that method?
- What did your group do to analyze the data, and why did you decide to do it that way?
- Is that the only way to interpret the results of your group's analysis? How do you know that your interpretation of the analysis is appropriate?
- Why did your group decide to present your evidence in that manner?
- What other claims did your group discuss before deciding on that one? Why did you abandon those alternative ideas?
- How confident are you that your group's claim is valid? What could you do to increase your confidence?

Once the argumentation session is complete, you will have a chance to meet with your group and revise your original argument. Your group might need to gather more data or design a way to test one or more alternative claims as part of this process. Remember, your goal at this stage of the investigation is to develop the most valid or acceptable answer to the research question!

Report

Once you have completed your research, you will need to prepare an *investigation report* that consists of three sections that provide answers to the following questions:

1. What question were you trying to answer and why?
2. What did you do during your investigation and why did you conduct your investigation in this way?
3. What is your argument?

Your report should answer these questions in two pages or less. The report must be typed and any diagrams, figures, or tables should be embedded into the document. Be sure to write in a persuasive style; you are trying to convince others that your claim is acceptable or valid!

Checkout Questions

Lab 28. Designing a Cold Pack: Which Salt Should Be Used to Make an Effective but Economical Cold Pack?

1. Describe the nature of the attractive forces that must be broken or disrupted when a polar solute is dissolved into a polar solvent.

2. Dissolving calcium chloride ($CaCl_2$) into water is an exothermic process. However, there are intermediate steps in the dissolving process that are endothermic. Use what you know about the process of dissolving to explain how it is possible for an exothermic process to actually involve some smaller endothermic processes.

3. Cultural and societal values have a great influence on science.

   a. I agree with this statement.
   b. I disagree with this statement.

   Explain your answer, using an example from your investigation about designing a cold pack.

4. Creativity is not valued in science; it is better to do things the way they have always been done.

   a. I agree with this statement.
   b. I disagree with this statement.

   Explain your answer, using an example from your investigation about designing a cold pack.

LAB 28

5. Understanding how energy and matter flow within systems is important in science, particularly when it comes to technological applications like designing a cold pack. Explain why understanding this aspect of science is important, using an example from your investigation.

6. Recognizing consistent patterns across several investigations is important in science. Explain the benefits of recognizing patterns, using an example from your investigation about designing a cold pack.

Lab Handout

Lab 29. Rate Laws: What Is the Rate Law for the Reaction Between Hydrochloric Acid and Sodium Thiosulfate?

Introduction

The *collision theory of reactions* suggests that the rate of a reaction depends on three important factors. The first is the number of collisions that take place between molecules during a reaction. This factor is important because molecules must collide with each other for a reaction to take place. The second factor is the average energy of these collisions. This factor is important because colliding molecules must have enough kinetic energy to overcome the repulsive and bonding forces of the reactants. The minimum amount of energy needed for the reactants to transform into products is called the *activation energy* of the reaction. The third, and final, factor is the orientation of the molecules at the time of the collision. This factor is important because the reactant molecules must collide with each other in a specific orientation for the atoms in these molecules to rearrange (see Figure L29.1). Chemists can therefore alter the rate of a specific chemical reaction by manipulating one or more of these factors inside the laboratory.

FIGURE L29.1
The collision theory of reactions

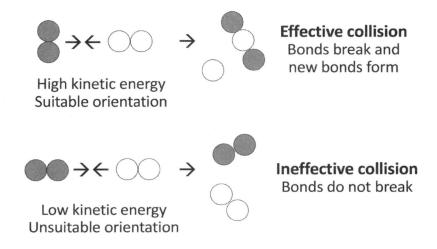

Chemists, for example, can change the temperature of the reactants to speed up or slow down a reaction. Temperature affects the rate of a reaction because the proportion of molecules with enough kinetic energy to overcome the activation energy barrier of a reaction goes up when the temperature increases and goes down when temperature decreases. Chemists can also change the concentration of the reactants to speed up or slow down the rate of a reaction. A change in reactant concentration changes the frequency of collisions between molecules but does not affect the likelihood that any given collision between any two molecules will be effective and result in a reaction. The likelihood of an effective collision, as noted earlier, depends on how much kinetic energy the molecules have and the orientation of the molecules when they collide with each other.

Chemists use an equation called a rate law to describe how the concentration of each reactant affects the overall reaction rate. For a general reaction of the form A + B → C, the rate law is written as

$$Rate = k[A]^n[B]^m$$

In this equation, k is the rate constant, [A] is the molar concentration of the reactant A, [B] is the molar concentration of the reactant B, and the exponents n and m define how the reaction rate depends on the concentration of each reactant. The rate constant for a reaction is affected by temperature but not by concentration. The n and m exponents are often described as the reaction order. In this example, the reaction is to the nth order of A and the mth order of B. Typical values of n and m are 0, 1, or 2.

The reaction order for each reactant in the rate law indicates how much the reaction rate will change in response to a change in the concentration of each reactant. If a reactant has an order of 0, then the reaction rate is not affected by the concentration of this reactant. Reactants with an order of 0 are therefore not included in the rate law. When a reactant has an order of 1, the change in reaction rate is directly proportional to the change in reactant concentration. The rate of a chemical reaction, for example, will double when the concentration of a first-order reactant is doubled and will triple when the concentration of the first-order reactant is tripled. When a reactant has an order of 2, the rate of reaction will change twice as much as the change in the reactant concentration. The rate of a chemical reaction, for example, will quadruple when the concentration of a second-order reactant is doubled.

It is important for chemists to understand how changing the concentration of one or more reactants will affect the overall rate of a specific chemical reaction so they can make a specific product in a safe and time-efficient manner. Unfortunately, chemists cannot determine the reaction order for each reactant in a specific reaction by simply looking at a balanced chemical equation; they must determine it through a process called the *differential method*. This method allows chemists to determine a rate law by varying the concentration of the reactants in a reaction in a systematic manner and then measuring how these changes affect the reaction rate. A description of the differential method can be found in the Reference Sheet (p. 251).

Rate Laws
What Is the Rate Law for the Reaction Between Hydrochloric Acid and Sodium Thiosulfate?

Your Task

Use the differential method to determine the rate law for the reaction between hydrochloric acid (HCl) and sodium thiosulfate ($Na_2S_2O_3$). The balanced chemical equation for this reaction is

$$2HCl(aq) + Na_2S_2O_3(aq) \rightarrow S(s) + SO_2(aq) + H_2O(l) + 2NaCl(aq)$$

The guiding question of this investigation is, **What is the rate law for the reaction between hydrochloric acid and sodium thiosulfate?**

Materials

You may use any of the following materials during your investigation:

Consumables	Equipment
• 1.0 M HCl • 0.3 M $Na_2S_2O_3$ • Distilled water	• 2 6-well reaction plates • 3 beakers (each 50 ml) • 3 graduated cylinders (each 10 ml) • 3 disposable pipettes • Stopwatch or timer

Safety Precautions

Follow all normal lab safety rules. Hydrochloric acid is moderately toxic by ingestion and inhalation, and it is also corrosive to the eyes and skin. Sodium thiosulfate is a body tissue irritant. The sulfur precipitate and aqueous sulfur dioxide, which are products of the reaction between hydrochloric acid and sodium thiosulfate, are skin and eye irritants. Your teacher will explain relevant and important information about working with the chemicals associated with this investigation. In addition, take the following safety precautions:

- Wear indirectly vented chemical-splash goggles and chemical-resistant gloves and apron while in the laboratory.
- Handle all glassware with care.
- Wash your hands with soap and water before leaving the laboratory.

Investigation Proposal Required?   ☐ Yes   ☐ No

Getting Started

The first step in the differential method is to carry out two experiments. The goal of the first experiment is to determine how a change in concentration of hydrochloric acid affects the reaction rate. The goal of the second experiment is to determine how a change in the concentration of sodium thiosulfate affects reaction rate. In each of your experiments, you will be able to measure reaction time by monitoring the appearance of the sulfur precipitate. The solution will change from clear to cloudy over time.

LAB 29

You will need to mix the reactants together in different concentrations in each of your experiments. You can change the concentration of each reactant by changing the amount of reactant you add to a well in a well plate. You can add between 1 and 4 ml of a reactant to a well, but the total volume of liquid added to each well should always equal 5 ml to keep volume constant across conditions. You can use water to bring the total volume of the liquid in a well up to 5 ml. For example, if you add 1 ml of HCl to 3 ml of $Na_2S_2O_3$, you will need to add 1 ml of water to bring the total volume of liquid in the well up to 5 ml. If you add 2 ml of HCl to 3 ml of $Na_2S_2O_3$, you will not need to add any water because the total volume of the liquid in the well is already 5 ml. Keep this information in mind as you design your two experiments. You will also need to determine what type of data you need to collect, how you will collect the data, and how you will analyze the data.

To determine *what type of data you need to collect*, think about the following questions:

- What types of measurements or observation will you need to make during each experiment?
- When will you need to make these measurements or observations?

To determine *how you will collect the data*, think about the following questions:

- What will serve as your independent variable in each experiment?
- What will serve as your dependent variable in each experiment?
- What comparisons will you need to make in each experiment?
- How will you determine when a reaction is finished?
- How will you hold other variables constant across each comparison?
- What will you do to reduce measurement error?
- How will you keep track of the data you collect and how will you organize it?

To determine *how you will analyze the data*, think about the following questions:

- How will you determine the concentration of each reactant at the beginning of the reaction?
- What type of calculations will you need to make?
- What type of graph could you create to help make sense of your data?

The second step in this investigation is to determine the rate law for this reaction. To determine the rate law, you will need to first calculate the reaction order for each reactant using the data you collected in each experiment. The Reference Sheet describes how to make these calculations. Next, you will need to calculate the value of k for the rate law. To calculate the rate constant, you simply need to substitute the appropriate values for the concentration of each reactant, the reaction order for each reactant, and the reaction rate into the rate law equation and solve for *k*.

Rate Laws
What Is the Rate Law for the Reaction Between Hydrochloric Acid and Sodium Thiosulfate?

The third and final step in the investigation is to test your rate law. To accomplish this task, you will need to determine if you can use your rate law to make accurate predictions about reaction times. If you can use the rate law you developed to make accurate predictions about the time it takes for the sulfur precipitate to appear when using different concentrations of hydrochloric acid and sodium thiosulfate, then you will have evidence that suggests that the rate law you developed is a valid one.

Connections to Crosscutting Concepts, the Nature of Science, and the Nature of Scientific Inquiry

As you work through your investigation, be sure to think about

- why it is important to look for and use proportional relationships;
- why it is important to track how energy and matter move into, out of, and within systems;
- the difference between observations and inferences in science; and
- the difference between laws and theories in science.

Initial Argument

Once your group has finished collecting and analyzing your data, you will need to develop an initial argument. Your argument must include a *claim*, which is your answer to the guiding question. Your argument must also include *evidence* in support of your claim. The evidence is your analysis of the data and your interpretation of what the analysis means. Finally, you must include a *justification* of the evidence in your argument. You will therefore need to use a scientific concept or principle to explain why the evidence that you decided to use is relevant and important. You will create your initial argument on a whiteboard. Your whiteboard must include all the information shown in Figure L29.2.

FIGURE L29.2
Argument presentation on a whiteboard

The Guiding Question:	
Our Claim:	
Our Evidence:	Our Justification of the Evidence:

Argumentation Session

The argumentation session allows all of the groups to share their arguments. One member of each group stays at the lab station to share that group's argument, while the other members of the group go to the other lab stations one at a time to listen to and critique the arguments developed by their classmates. The goal of the argumentation session is not to convince others that your argument is the best one; rather, the goal is to identify errors or instances of faulty reasoning in the initial arguments so these mistakes can be fixed. You will therefore need to evaluate the content of the claim, the quality of the evidence used

LAB 29

to support the claim, and the strength of the justification of the evidence included in each argument that you see. To critique an argument, you might need more information than what is included on the whiteboard. You might, therefore, need to ask the presenter one or more follow-up questions, such as:

- How did your group collect the data? Why did you use that method?
- What did your group do to make sure the data you collected are reliable? What did you do to decrease measurement error?
- What did your group do to analyze the data, and why did you decide to do it that way? Did you check your calculations?
- Is that the only way to interpret the results of your group's analysis? How do you know that your interpretation of the analysis is appropriate?
- Why did your group decide to present your evidence in that manner?
- What other claims did your group discuss before deciding on that one? Why did you abandon those alternative ideas?
- How confident are you that your group's claim is valid? What could you do to increase your confidence?

Once the argumentation session is complete, you will have a chance to meet with your group and revise your original argument. Your group might need to gather more data or design a way to test one or more alternative claims as part of this process. Remember, your goal at this stage of the investigation is to develop the most valid or acceptable answer to the research question!

Report

Once you have completed your research, you will need to prepare an *investigation report* that consists of three sections that provide answers to the following questions:

1. What question were you trying to answer and why?
2. What did you do during your investigation and why did you conduct your investigation in this way?
3. What is your argument?

Your report should answer these questions in two pages or less. The report must be typed and any diagrams, figures, or tables should be embedded into the document. Be sure to write in a persuasive style; you are trying to convince others that your claim is acceptable or valid!

Rate Laws
What Is the Rate Law for the Reaction Between Hydrochloric Acid and Sodium Thiosulfate?

Reference Sheet

The Differential Method for Determining a Rate Law

The following procedure can be used to determine the reaction order and the rate law for a reaction of the form A + B → C.

1. The concentration of reactant A is held constant while the concentration of reactant B is varied, and the reaction time is measured in each condition. Then the concentration of reactant B is held constant while the concentration of reactant A is varied, and the reaction time is measured in each condition. The rate for each condition is then calculated by taking the inverse of the reaction time. Take the following data as an example:

Condition	[A] (moles)	[B] (moles)	Reaction time (sec)	Rate (1/time)
1	0.02	0.01	500	0.002
2	0.01	0.01	1,000	0.001
3	0.01	0.02	250	0.004

2. Determine the reaction order with respect to A (n) by comparing conditions 1 and 2. To do this, use the following equation:

$$Rate_{condition1}/Rate_{condition2} = ([A]_{condition1}/[A]_{condition2})^n$$

In this example,

$$0.002/0.001 = (0.02/0.01)^n$$

$$2 = 2^n$$

$$n = 1$$

3. Determine the reaction order with respect to B (m) by comparing conditions 2 and 3. To do this, use the following equation:

$$Rate_{condition3}/Rate_{condition2} = ([B]_{condition3}/[B]_{condition2})^m$$

In this example,

$$0.004/0.001 = (0.02/0.01)^m$$

$$4 = 2^m$$

$$m = 2$$

4. Write out the full rate law.

In this example,

$$Rate = k[A][B]^2$$

Checkout Questions

Lab 29. Rate Laws: What Is the Rate Law for the Reaction Between Hydrochloric Acid and Sodium Thiosulfate?

Use the following information to answer questions 1–4.

Iodide ions (I^-) react with persulfate ions ($S_2O_8^{2-}$) as follows:

$$2I^-(aq) + S_2O_8^{2-}(aq) \rightarrow I_2(aq) + 2SO_4^{2-}(aq)$$

The following rate data were collected by measuring the time required for the appearance of a dark blue color due to the interaction of iodine and starch.

Condition	[I^-] (moles)	[$S_2O_8^{2-}$] (moles)	Reaction time (sec)
1	0.04	0.04	270
2	0.08	0.04	138
3	0.04	0.08	142

1. What is the reaction rate for each condition?

2. What is the order of reaction for the iodide ions?

Rate Laws
What Is the Rate Law for the Reaction Between Hydrochloric Acid and Sodium Thiosulfate?

3. What is the order of reaction for the persulfate ions?

4. What is the rate law for this reaction?

5. "The reactant molecules collide during the reaction" is an observation.

   a. I agree with this statement.
   b. I disagree with this statement.

   Explain your answer, using an example from your investigation about rate laws.

LAB 29

6. Theories can turn into laws.

   a. I agree with this statement.
   b. I disagree with this statement.

   Explain your answer, using an example from your investigation about rate laws.

7. Scientists often need to look for and use proportional relationships during an investigation. Explain why this is important, using an example from your investigation about rate laws.

8. Scientists often need to track how matter and energy move into, out of, and within a system. Explain why this is important, using an example from your investigation about rate laws.

Equilibrium Constant and Temperature
How Does a Change in Temperature Affect the Value of the Equilibrium Constant for an Exothermic Reaction?

Lab Handout

Lab 30. Equilibrium Constant and Temperature: How Does a Change in Temperature Affect the Value of the Equilibrium Constant for an Exothermic Reaction?

Introduction

Chemical equilibrium is defined as the point in a reaction where the rate at which reactants transform into products is equal to the rate at which products revert back into reactants. When a reaction is in equilibrium, the concentration of the products and reactants is constant or stable. At this point, there is no further net change in the amounts of reactants or products unless the system is disturbed in some manner. The *equilibrium constant* provides a mathematical description of the equilibrium state for any reversible chemical reaction. To illustrate, consider the following general equation for a reversible reaction:

$$aA + bB \rightleftarrows cC + dD$$

The equation for calculating the equilibrium constant, K_{eq}, for this general reaction is provided below. The square brackets refer to the molar concentrations of each substance at equilibrium. The exponents are the stoichiometric coefficients of each substance found in the balanced chemical equation.

$$K_{eq} = \frac{[C]^c[D]^d}{[A]^a[B]^b}$$

The equilibrium constant describes the proportional relationship that exists between the concentration of the reactants and the concentration of the products for a specific chemical reaction when the reaction is in a state of equilibrium. The actual concentrations of the reactants and products that are present in the system at equilibrium will depend on the initial amounts of the reactants that were used at the beginning of the reaction and any extra reactants or products that were added to the system after the reaction started. The concentration ratio of products to reactants described by the equilibrium constant, however, will always be the same as long as the system is in equilibrium and the temperature of the system does not change.

The equilibrium constant is useful because it allows chemists to determine the product-to-reactant concentration ratio that will be present in the reaction mixture at equilibrium before the reaction begins. When $K_{eq} > 1$, the concentration of the products in the system will be greater than the concentration of the reactants. When $K_{eq} < 1$, the concentration of products will be less than the concentration of the reactants. Finally, when $K_{eq} = 1$, the concentration of products and reactants will be equal. A reaction with a large K_{eq} value, as a result, will have a greater product-to-reactant concentration ratio at equilibrium than a reaction with a smaller value.

The equilibrium constant of a reaction will change when the temperature changes. A positive shift in the K_{eq} will cause the product-to-reactant concentration ratio at equilibrium to increase, whereas a negative shift will result in a decrease in the concentration ratio. How the K_{eq} of a reaction will change in response to a change in temperature, however, is not uniform. The change in the K_{eq} of a reaction depends on the direction and magnitude of the temperature change. It will also depend on whether the reaction is exothermic or endothermic.

To control the amount of product or reactant present at the equilibrium point of a reaction in a closed system, chemists need to understand how the equilibrium constant will shift in response to changes in temperature for different types of reactions. You will therefore determine the equilibrium constant for an exothermic reaction and then explore how increases and decreases in temperature change the value of the equilibrium constant for this reaction. You will then develop a rule that you can use to predict how a change in temperature will affect the value of the equilibrium constant for other exothermic reactions.

Your Task

Determine the equilibrium constant for the reaction between iron(III) nitrate and potassium thiocyanate. Then determine how the equilibrium constant for this exothermic reaction is affected by a change in temperature. Once you understand how the equilibrium constant for this reaction changes in response to a change in temperature, you will need to develop a rule that you can use to predict how the equilibrium constant of other exothermic reactions will change in a response to a temperature change.

The guiding question of this investigation is, **How does a change in temperature affect the value of the equilibrium constant for an exothermic reaction?**

Materials

You may use any of the following materials during your investigation:

Consumables	Equipment
• 0.002 M iron(III) nitrate, $Fe(NO_3)_3$ • 0.200 M iron(III) nitrate, $Fe(NO_3)_3$ • 0.002 M potassium thiocyanate, KSCN • Distilled water • Ice	• Colorimeter sensor • Sensor interface • 4 cuvettes • 12 test tubes • Test tube rack • 3 serological pipettes (each 5 or 10 ml) • Pipette bulb • Stirring rod • 6 beakers (each 50 ml) • 2 beakers (each 250 ml, for water baths) • Hot plate • Thermometer

Equilibrium Constant and Temperature
How Does a Change in Temperature Affect the Value of the Equilibrium Constant for an Exothermic Reaction?

Safety Precautions

Follow all normal lab safety rules. Potassium thiocyanate is toxic by ingestion. Iron(III) nitrate solution contains 1 M nitric acid and is a corrosive liquid; it will also stain clothes and skin. Your teacher will explain relevant and important information about working with the chemicals associated with this investigation. In addition, take the following safety precautions:

- Wear indirectly vented chemical-splash goggles and chemical-resistant gloves and apron while in the laboratory.
- Use caution when working with hot plates because they can burn skin. Hot plates also need to be kept away from water and other liquids.
- Handle all glassware (including thermometers) with care.
- Wash hands with soap and water before leaving the laboratory.

Investigation Proposal Required?   ☐ Yes   ☐ No

Getting Started

The first step in this investigation is to determine the equilibrium constant for the reaction between iron(III) nitrate and potassium thiocyanate at room temperature. Iron(III) ions react with thiocyanate ions to form FeSCN^{2+} complex ions according to the following reaction:

$$Fe^{3+}(aq) + SCN^-(aq) \leftrightarrows FeSCN^{2+}(aq)$$

Yellow   Colorless   Orange-Red

The equilibrium constant expression for this reaction is

$$K_{eq} = \frac{[FeSCN^{2+}]}{[Fe^{3+}][SCN^-]}$$

You can determine the value of K_{eq} by mixing solutions with known concentrations of Fe^{3+} and SCN$^-$ and then measuring the concentration of the FeSCN^{2+} ions in the mixture once the reaction is at equilibrium. The equilibrium concentration of the FeSCN^{2+} ions in the solution can be determined by measuring the absorbance of the solution using a colorimeter. This is possible because the FeSCN^{2+} ions produce a red color and the amount of light absorbed by the solution is directly proportional to the concentration of the FeSCN^{2+} ions in it. You can, as a result, determine the FeSCN^{2+} concentration of any solution by simply comparing the absorbance of that solution with the absorbance of a solution with a known FeSCN^{2+} concentration (called a standard solution).

You will need to make a standard solution and five different test solutions to determine the equilibrium constant for the reaction between iron(III) nitrate and potassium thiocyanate at room temperature. Prepare the standard solution and five test solutions as described in Table L30.1 (p. 258).

TABLE L30.1

Components of the standard and test solutions

Sample	Reactants (ml)			
	0.200 M Fe(NO$_3$)$_3$	0.002 M Fe(NO$_3$)$_3$	0.002 M KSCN	Distilled water
Standard solution	9.00	0.00	1.00	0.00
Test solution 1	0.00	5.00	1.00	4.00
Test solution 2	0.00	5.00	2.00	3.00
Test solution 3	0.00	5.00	3.00	2.00
Test solution 4	0.00	5.00	4.00	1.00
Test solution 5	0.00	5.00	5.00	0.00

Once you have your solutions prepared, you can measure the absorbance of each one. Your teacher will show you how to use the calorimeter to measure the absorbance of the solutions. You will need to determine the concentration of FeSCN^{2+} in each test solution and then use this information to calculate an average equilibrium constant for the reaction. To accomplish this task, follow the procedure below:

1. Calculate the concentration of the Fe^{3+} and SCN$^-$ ions in the standard solution and the five test solutions using the dilution equation ($M_1V_1 = M_2V_2$).

2. Calculate the concentration of the FeSCN^{2+} ions in the standard solution and each test solution at equilibrium using the following equation:

$$[FeSCN^{2+}]_{Equilibrium} = \frac{A_{TestSolution}}{A_{StandardSolution}} \times [FeSCN^{2+}]_{Standard}$$

Assume the concentration of the FeSCN^{2+} in the standard solution at equilibrium is equal to the concentration of the SCN$^-$ ions in the standard solution at the start of the reaction. You can make this assumption because all of the SCN$^-$ ions should have been converted to FeSCN^{2+} ions due to the large amount of Fe^{3+} that was added to the standard solution.

3. Calculate the equilibrium concentration of Fe^{3+} ions in each test solution by subtracting the equilibrium concentration of FeSCN^{2+} from the initial concentration of Fe^{3+} ions using the equation

$$[Fe^{3+}]_{TestSolutionEquilibrium} = [Fe^{3+}]_{TestSolutionInitial} - [FeSCN^{2+}]_{TestSolution}$$

Equilibrium Constant and Temperature
How Does a Change in Temperature Affect the Value of the Equilibrium Constant for an Exothermic Reaction?

4. Calculate the equilibrium concentration of SCN^- ions in each test solution by subtracting the equilibrium concentration of $FeSCN^{2+}$ from the initial concentration of SCN^- ions using the equation

$$[SCN^-]_{TestSolutionEquilibrium} = [SCN^-]_{TestSolutionInitial} - [FeSCN^{2+}]_{TestSolution}$$

5. Calculate the value of the equilibrium constant for the five test solutions using the equation

$$K_{eq} = \frac{[FeSCN^{2+}]}{[Fe^{3+}][SCN^-]}$$

6. Calculate the average equilibrium constant for the reaction.

The second step in your investigation is to conduct an experiment to determine how a change in temperature affects the equilibrium constant for the reaction between iron(III) nitrate and potassium thiocyanate. To conduct this experiment, you must determine what type of data you need to collect, how you will collect the data, and how you will analyze the data.

To determine *what type of data you need to collect*, think about the following questions:

- What type of measurements or observations will you need to record during each experiment?
- When will you need to make these measurements or observations?

To determine *how you will collect the data*, think about the following questions:

- What will serve as your independent variable?
- What types of comparisons will you need to make?
- How will you change the temperature of the reaction?
- What will you do to reduce measurement error?
- How will you keep track of the data you collect and how will you organize it?

To determine *how you will analyze the data*, think about the following questions:

- What type of calculations will you need to make?
- What type of graph could you create to help make sense of your data?

The last step in this investigation is to develop a rule that you can use to predict how the equilibrium constant of other exothermic reactions will change in a response to a temperature change. This rule will serve as your answer to the guiding question of this investigation.

LAB 30

Connections to Crosscutting Concepts, the Nature of Science, and the Nature of Scientific Inquiry

As you work through your investigation, be sure to think about

- how scientists must define the system they are studying and then use models to understand it,
- why it is important to understand what makes a system stable or unstable and what controls the rates of change in a system,
- the importance of imagination and creativity in science, and
- the role of experiments in science.

Initial Argument

Once your group has finished collecting and analyzing your data, you will need to develop an initial argument. Your argument must include a *claim*, which is your answer to the guiding question. Your argument must also include *evidence* in support of your claim. The evidence is your analysis of the data and your interpretation of what the analysis means. Finally, you must include a *justification* of the evidence in your argument. You will therefore need to use a scientific concept or principle to explain why the evidence that you decided to use is relevant and important. You will create your initial argument on a whiteboard. Your whiteboard must include all the information shown in Figure L30.1.

FIGURE L30.1
Argument presentation on a whiteboard

The Guiding Question:	
Our Claim:	
Our Evidence:	Our Justification of the Evidence:

Argumentation Session

The argumentation session allows all of the groups to share their arguments. One member of each group stays at the lab station to share that group's argument, while the other members of the group go to the other lab stations one at a time to listen to and critique the arguments developed by their classmates. The goal of the argumentation session is not to convince others that your argument is the best one; rather, the goal is to identify errors or instances of faulty reasoning in the initial arguments so these mistakes can be fixed. You will therefore need to evaluate the content of the claim, the quality of the evidence used to support the claim, and the strength of the justification of the evidence included in each argument that you see. To critique an argument, you might need more information than what is included on the whiteboard. You might, therefore, need to ask the presenter one or more follow-up questions, such as:

- How did your group collect the data? Why did you use that method?
- What did your group do to make sure the data you collected are reliable? What did you do to decrease measurement error?

Equilibrium Constant and Temperature
How Does a Change in Temperature Affect the Value of the Equilibrium Constant for an Exothermic Reaction?

- What did your group do to analyze the data, and why did you decide to do it that way? Did you check your calculations?
- Is that the only way to interpret the results of your group's analysis? How do you know that your interpretation of the analysis is appropriate?
- Why did your group decide to present your evidence in that manner?
- What other claims did your group discuss before deciding on that one? Why did you abandon those alternative ideas?
- How confident are you that your group's claim is valid? What could you do to increase your confidence?

Once the argumentation session is complete, you will have a chance to meet with your group and revise your original argument. Your group might need to gather more data or design a way to test one or more alternative claims as part of this process. Remember, your goal at this stage of the investigation is to develop the most valid or acceptable answer to the research question!

Report

Once you have completed your research, you will need to prepare an *investigation report* that consists of three sections that provide answers to the following questions:

1. What question were you trying to answer and why?
2. What did you do during your investigation and why did you conduct your investigation in this way?
3. What is your argument?

Your report should answer these questions in two pages or less. The report must be typed and any diagrams, figures, or tables should be embedded into the document. Be sure to write in a persuasive style; you are trying to convince others that your claim is acceptable or valid!

LAB 30

Checkout Questions

Lab 30. Equilibrium Constant and Temperature: How Does a Change in Temperature Affect the Value of the Equilibrium Constant for an Exothermic Reaction?

1. Determine the equilibrium constant, K_{eq}, for the following chemical reaction at equilibrium if the molar concentrations of the molecules are 0.20 M H_2, 0.10 M NO, 0.20 M H_2O, and 0.10M N_2:

$$2H_2(g) + 2NO(g) \rightarrow 2H_2O(g) + N_2(g)$$

Is there a greater concentration of total products or reactants in this equilibrium situation?

   a. Greater concentration of total products
   b. Greater concentration of total reactants

How do you know?

2. In the following reaction, the temperature is increased and the K_{eq} value decreases from 0.75 to 0.55:

$$N_2(g) + 3H_2 \leftrightarrow 2NH_3(g)$$

Equilibrium Constant and Temperature
How Does a Change in Temperature Affect the Value of the Equilibrium Constant for an Exothermic Reaction?

What kind of reaction is this?

  a. Exothermic
  b. Endothermic

How do you know?

3. In the following reaction, the enthalpy of reaction is ΔH = –92.5 kJ and there is an increase in temperature:

$$PCl_3(g) + Cl_2(g) \leftrightarrow PCl_5(g)$$

How will the equilibrium constant shift?

  a. It will increase.
  b. It will decrease.

How do you know?

4. Scientists use experiments to prove that ideas are true.

  a. I agree with this statement.
  b. I disagree with this statement.

Explain your answer, using an example from your investigation about equilibrium constant and temperature.

LAB 30

5. Scientists need to be creative and have a good imagination to excel in science.

    a. I agree with this statement.
    b. I disagree with this statement.

    Explain your answer, using an example from your investigation about equilibrium constant and temperature.

6. An important goal in science is to understand what types of disturbances can make a system unstable and how a system will respond to a disturbance. Explain why this is important, using an example from your investigation about equilibrium constant and temperature.

7. Scientists often need to define a system under study and then use or develop a model to help them understand it. Explain why this is important, using an example from your investigation about equilibrium constant and temperature.

IMAGE CREDITS

LAB 1
Figure L1.1: PhET Interactive Simulations, University of Colorado, http://phet.colorado.edu; http://phet.colorado.edu/en/simulation/molecule-polarity

LAB 2
Figure L2.1: Authors

Figure L2.2: PhET Interactive Simulations, University of Colorado, http://phet.colorado.edu; http://phet.colorado.edu/en/simulation/molecule-shapes

LAB 4
Figure L4.1: PhET Interactive Simulations, University of Colorado, http://phet.colorado.edu; http://phet.colorado.edu/en/simulation/molarity

LAB 5
Figure L5.1: Authors

LAB 6
Figure L6.1: Authors

Figure L6.2: Authors

LAB 7
Figure L7.1: Chris King, Wikimedia Commons, CC BY-SA 3.0, GFDL 1.2. http://commons.wikimedia.org/wiki/File:Atomic_%26_ionic_radii.svg

LAB 8
Figure L8.1: Authors

Figure L8.2: Authors

LAB 10
Figure L10.1: Left: User:LHcheM, Wikimedia Commons, CC BY-SA 3.0, GFDL 1.2. http://commons.wikimedia.org/wiki/File:Sample_of_Methyl_Blue.jpg; Right: User:LHcheM, Wikimedia Commons, CC BY-SA 3.0, GFDL 1.2. http://commons.wikimedia.org/wiki/File:Methyl_Blue_aqueous_solution.jpg

LAB 11
Figure L11.1: CK-12 Foundation, Wikimedia Commons, CC BY-SA 3.0. http://commons.wikimedia.org/wiki/File:Px_py_pz_orbitals.png

Figure L11.2: Authors

Figure L11.3: Authors

LAB 12
Figure L12.1.: User:Sven, Wikimedia Commons, CC BY-SA 3.0, GFDL 1.2. http://commons.wikimedia.org/wiki/File:S-p-Orbitals.svg

Figure L12.2: CK-12 Foundation, Wikimedia Commons, Public domain. http://commons.wikimedia.org/wiki/File:Electron_configuration_iron.svg

Figure L12.3: Will Thomas Jr.

LAB 13
Figure L13.1: Wikimedia Commons, Public domain. http://commons.wikimedia.org/wiki/File:Mendeleev%27s_1869_periodic_table.png

LAB 17
Figure L17.1: Authors

Figure L17.2: Authors

Checkout Questions: Authors

Image Credits

LAB 19
Figure L19.1: Authors

Figure L19.2: Authors

LAB 20
Figure L20.1: Authors

Figure L20.2: Authors

LAB 22
Figure L22.1: Authors

LAB 24
Figure L24.1: User:PRHaney, Wikimedia Commons, CC BY-SA 3.0. *http://commons.wikimedia.org/wiki/File:Lead_%28II%29_iodide_precipitating_out_of_solution.JPG*

LAB 25
Figure L25.1: User:Liquid_2003, Wikimedia Commons, CC BY 2.0. *http://commons.wikimedia.org/wiki/File:Titrage.svg*

LAB 27
Figure L27.1: Authors

LAB 28
Figure L28.1: Authors

LAB 29
Figure L29.1: Authors